SpringerBriefs in Genetics

For further volumes:
http://www.springer.com/series/8923

Xuhua Xia

Comparative Genomics

 Springer

Xuhua Xia
Department of Biology
University of Ottawa
Ottawa, ON
Canada

ISSN 2191-5563   ISSN 2191-5571   (electronic)
ISBN 978-3-642-37145-5   · ISBN 978-3-642-37146-2   (eBook)
DOI 10.1007/978-3-642-37146-2
Springer Heidelberg New York Dordrecht London

Library of Congress Control Number: 2013936000

Printed on acid-free paper

Springer is part of Springer Science+Business Media (www.springer.com)

Preface

This book on comparative genomics was written for early researchers (advanced undergraduate students, postgraduates, and postdoctoral fellows). Well-established biologists should leave it alone—it is not intended to impress them.

What is comparative genomics? Before a proper definition can be put forward, we need to recognize that a genome has many primary features such as the genomic sequence, strand asymmetry, genes, gene order, regulatory motifs, and genomic structural landmarks that can be recognized or modified by cellular components with functional implications, etc. A genome also has secondary features such as the dynamic transcriptome, proteome, codon–anticodon adaptation, functional association of genes, and gene interaction networks. Comparative genomics is a branch of genomics that aims to (1) characterize the similarity and differences in genomic features and trace their gain and loss along different evolutionary lineages, (2) understand the evolutionary forces such as mutation and selection that govern the changes of these genomic features, and (3) find out how genomic evolution can help us battle diseases, restore environmental health, make money, etc.

It is better to illustrate this with an example. Suppose we have a set of bacterial genomes, with Genome A missing genes for lactose metabolism in contrast to all closely related genomes that still carry the genes. We may reasonably infer that the genes were lost in the lineage leading to Genome A. Suppose we further find that the organism carrying Genome A has inhabited an environment that is constantly lactose-free (I, as well as some of my Chinese, Finnish and German colleagues, would love to have such an environment), then we can infer that genetic alterations to the lactose-metabolizing genes are essentially neutral for the carrier of Genome A, with no functional consequence for losing the gene. Through a phylogeny-based analysis, we may find that lactose-free environment is strongly associated with the loss of lactose-metabolizing genes. If we further find that the set of genes are either strongly conserved in evolutionary lineages requiring lactose metabolism or degraded by accumulated mutations in those living in lactose-free environment, we can infer that the genes are strongly associated only for the lactose-metabolizing function. In contrast, if we find that the set of genes are still strongly conserved in lineages inhabiting lactose-free environment for a long time, then the genes may have functions other than lactose metabolism.

What basic knowledge do we need to do research in comparative genomics? The most fundamental feature of a single genome is its nucleotide sequence, and the most fundamental feature shared among a set of genomes is coancestry, or shared homology. These immediately bring into our mind the necessity of sequence-related computational tools such as sequence alignment and molecular phylogeny. For this reason, some literacy in computation and mathematics/statistics is assumed.

Much of the comparative genomics is done by genomic comparison against genomes of model organisms. Consequently, it is of tremendous value to gain a good understanding of molecular biology of some model organisms such as *Escherichia coli*, *Bacillus subtilis*, *Mycoplasma genitalium*, *Chlamydomonas reinhardtii*, *Arabidopsis thaliana*, *Saccharomyces cerevisiae*, *Caenorhabditis elegans*, *Drosophila melanogaster*, *Ciona intestinalis*, *Danio rerio*, *Takifugu rubripes*, *Xenopus laevis*, *Gallus gallus*, *Mus musculus* and, of course, *Homo sapiens*. For an evolutionary biologist, it is a great comfort to see such a diverse array of model organisms, especially for those who have lived through the bygone era dominated by the dogmatic assertion that "What is true in *E. coli* is also true in the elephant".

What about viruses? Can one do research in comparative genomics of viral genomes? The main difficulty with viral genomes is that viral lineages are often so diverse that they do not share any detectable homology. So comparative genomics is typically limited to closely related lineages such as among different subtypes of influenza viruses or among HIV/SIV viruses. However, lack of homology does not preclude one extremely important aspect of evolutionary studies, i.e., the study of convergent evolution. Diverse bacteriophage lineages can parasitize the same host and serve as a fertile ground for studying convergent evolution in response to the same intracellular environment of the host. However, it is the demonstration of functional equivalence, instead of homology, of the genes that is at the center of lime light in the study of convergent evolution in comparative viral genomics.

Comparative genomic research should be guided by the conceptual framework of evolutionary biology, so readers are assumed to have read something Darwinian. There are two most fundamental problems in evolutionary biology. The first is the origin and maintenance of new features and new species. There is no better way to address this question than comparative genomics, where the gain and loss of functional genes, as well as modification of a gene to gain a new function, can often be unequivocally identified from a set of related genomes. Many bacterial species are competent in pick up environmental DNA segments and integrate them into their genomes. Some of these DNA segments contain functional genes, leading to inheritance of the newly "acquired characters" and changes in subsequent evolutionary trajectories.

The second fundamental problem in evolutionary biology is the establishment of the links among genotype, phenotype, and environment. The greatest stumbling block to this line of enquiry has been the characterization of the genotype. This block is essentially non-existent when we have all the genomes and can characterize various aspects of the genotype, e.g., the presence/absence of a set of genes. We can then use phylogeny-based methods to systematically characterize the association between this matrix of genotypes and the matrix of phenotypes or the matrix of environmental factors.

The diverse genomes we see today did not originate independently, but represent products of descent with modification. This has fundamental implications on the methodology in comparative genomics. A good phylogeny is typically required for any comparative genomic study involving more than two genomes. The reader is therefore assumed to have gained basic understanding of phylogenetics.

Many examples of comparative genomic research are illustrated throughout the book. The first chapter includes many small-scale research examples, while the second chapter is heavy with large-scale studies and their associated statistical methods, in particular the comparative methods involving both continuous and discrete variables. The effort to develop phylogeny-based comparative methods was initiated by Joe Felsenstein and subsequently further developed and promoted by Paul Harvey and Mark Pagel. I numerically illustrated these methods in such a way that researchers with basic statistical and programing skills can include these methods in their programs. It should also facilitate further development of the methods by people well-versed in stochastic processes. The third chapter presents frequently used methods for detecting viral recombination.

The comparative approach has gone way beyond biology. For example, social scientists have characterized "phenotypes" of different forms of government and how much of the "phenotypic" differences can be attributed to historical inertia and environmental and cultural determinants. From a social biogeographic point of view, there are two possibilities for why Government Form A (GF_A) is found in Area X but GF_B is found in Area Y. First, GF_A is "good" for people in Area X and "bad" for people in Area Y. Likewise, GF_B is "good" for people in Area Y but "bad" for people in Area X. In this case, we should leave these people alone. Second, GF_A is "better" than GF_B in both areas but has never got a chance to be practised by people in Area Y. In this case, we might try to persuade people in Area Y to practise GF_A. Phylogeny-based methods can help us discriminate between the two possibilities, although some politicians and religious leaders have long settled for the second possibility, i.e., one particular GF or religion is better than all alternatives and should be promoted and practised everywhere in the world.

This book is not on democracy or religion, and is not good for everyone. In fact, book authors universally acknowledge the truth that a book is never good for everyone. For this reason, many authors are profusely apologetic in the preface, although there are also a few courageous ones who simply stated "Please read the book". I do not want to be apologetic and obviously do not want to draw reader's attention to problems in my book, but feel that I have to list a few things below just to conform to the convention.

First, this book does not cover all aspects of comparative genomics. In particular, it does not cover any aspect of genome rearrangement, for three reasons. First, many books entitled "Comparative Genomics" include extensive coverage of genome rearrangement. Second, most genes in eukaryotes and operons in prokaryotes appear to function well without being constrained by their location in the genome. Third, I myself do not work on genome rearrangement, which is my strongest justification for the omission. I do not think that anyone wants to read a professional book, or even part of it, written by a layperson.

Second, do not be infuriated when you find your important works not cited in the book because this book has a mandate to be brief. If you keep up your good work, readers of the book will discover you sooner or later. You would be a modern Mendel if you get rediscovered by three separate investigators, which perhaps is not a bad thing after all.

Third, I am a Chinese, and English is not my mother tongue. If you come across a grammatical error, please do not immediately shred the book or angrily demand refund. Let me see if I can squeeze a smile out of you by sharing a little story of me. The textbook of English during my undergraduate years in China typically had a list of new English words/phrases and their Chinese equivalents side by side. "Should" and "to be supposed to" happened to have the same Chinese equivalent that means "should", and I had since considered "should" and "to be supposed to" as synonymous. Then there came a time when I was doing my graduate research in a field station with a group of other Canadian students. I typically would wash dishes because others did the cooking which took much more time and energy. Once my fellow students suggested that I should share the dishwashing with others, and I wanted to say "I should wash the dishes" because others did the cooking. But then I thought that "to be supposed to" seemed much more grandiose than the plain "should". So I replied that "I am supposed to wash the dishes", privately thinking that they would be really impressed by my command of English. The resulting behaviour of my Canadian fellow students puzzled me for a whole field season, and I wrote home that "culture shock" was so real and that Canadians could truly be weird and unpredictable.

I hope that this book will not create many "weird and unpredictable" readers.

Acknowledgments

An experienced publisher once pointed me to a few examples of "effective use" of acknowledgment, each with an impressive list of well-known scientists, tactfully acknowledged to boost the reputation of the book author. The practice reminded me of some recent scientific conferences each with a list of 8–11 Nobel laureates as session chairs or keynote speakers. A journal would not have legitimacy if it does not have a list of silverbacks in the editorial board, even though some of the silverbacks are never involved in the manuscript-screening process. A person's worth is often evaluated by the number of "like" in social networks. We are entering a world in which a masterpiece in art is no longer evaluated on its own merit, but on whether it features gold-plated frame or displayed in a prominent location in a museum or gallery!

Should I mould a few famous names into a gold-plated frame for my limited painting of comparative genomics? I did have the good fortune of being associated with a number of silverbacks. Some helped me to switch to molecular evolution and phylogenetics when I was forced to switch fields because of severe allergies toward rodents that I used to study. Some offered me their books as gifts that inspired me and cultivated in my mind a strong desire to produce something similar. Some donated their previous field data or bacterial strains that led to results included in this book. Some have commented much of the book and corrected errors in the second chapter of this book. However, there are also little known people, but much greater in number, who have helped me and supported me in various ways during the writing process. If the "effective use" of Acknowledgement implies the exclusion of little known names, then let me engrave all these names in my heart without mentioning any here. I think that they would all like it this way.

But some explicit acknowledgments are absolutely essential—there would be serious repercussions if I did not. Scientists, just as religious monks, need patronage to carry out their daily routines and rituals. Without generous patronage, there would be neither religious freedom nor academic freedom. So here goes my acknowledgment to funding agencies: NSERC (Discovery Grant) and CAS/SAFEA (International Partnership Program for Creative Research Teams). While the money has never been sufficient for research, it is perhaps worth as much as a gold-plated frame for decorating the book.

I should also thank Evelyn Best who encouraged me to write this book as an expansion of a previous book chapter. I was initially reluctant because the word "expansion" reminds me of software bloating. To paraphrase Joe Armstrong (creator of Erlang), when a reader asks for only a banana, should I give him a gorilla holding a banana or even an entire jungle? However, I soon realized that the banana alone does not make a healthy meal. Hence this book, with some additional berries, but no gorilla or jungle in it.

My limited command of the English language becomes particularly acute when I come to express my appreciation for my wife (Zheng) and my children. They are a miracle to me. The arrow of time has brought so much wonderful transformation to our little ones and created so many memorable moments. By just looking at them, I am convinced that the world after me will be much nicer, gentler, and smarter. May they grow up and enjoy reading this book!

Contents

Contents

Chapter 1
What is Comparative Genomics?

Some scientists are visionary and can envision the theoretical foundation and experimental methodology of a new branch of science long before it takes any concrete shape. However, most scientists are just classifiers. When they see colleagues engage in novel activities such as catching flies, killing mice, chasing elephants in Africa and mounting whale specimen for museums, they would create a container labelled "zoology" and dump all these activities into it. Similarly, all those activities such as climbing trees, picking flowers, growing *Arabidopsis thaliana* and maintaining greenhouses are boxed together as botany. One former colleague of mine claimed that the only exception to this naming convension involves those studies of feces in hospitals—they are lumped together as microbiology instead of a potentially more descriptive name.

Then what is comparative genomics? Following the convention of classification, we simply define comparative genomics as the collection of all research activities that derive biological insights by comparing genomic features. A genome has many features such as the genomic sequence, strand asymmetry, genes, gene order, regulatory sequences, genomic structural landmarks that can be recognized or modified by cellular components with functional implications, etc. Comparative genomics is a branch of genomics that aims to (1) characterize the similarity and differences in genomic features and trace their origin, change and loss along different evolutionary lineages, (2) understand the evolutionary forces such as mutation, recombination, lateral gene transfer, and selection (mediated by abiotic environment such as temperature, food, and pH and biotic factors such as host, parasite, and competitors) that govern the changes of these genomic features, and (3) find out how genomic evolution can help us battle diseases by developing personalized medicine, improve environmental health, restore sustainable development, etc.

The development of comparative genomics predates the availability of genomic sequences. It has long been known that organisms are genetically related, with many homologous genes sharing similar functions among diverse organisms.

For example, the yeast *IRA2* gene is homologous to the human *NF1* gene, and the functional equivalence of the two genes was demonstrated by the yeast *IRA2* mutant being rescued by the human *NF1* gene (Ballester et al. 1990). This suggests the possibility that simple genomes can be used as a model to study complicated genomes. A multitude of such demonstrations of functional equivalence of homologous genes across diverse organisms has led to the dogmatic assertion that what is true in *E. coli* is also true in the elephant (attributed to Jacques Monod, Jacob 1988, p. 290).

It is the realization that what is true in *E. coli* is often not true in the elephant that has brought comparative genomics into the proper evolutionary context with the concept of phylogenetic controls. This is best illustrated by a simple example. Suppose we compare two Dodge Caravans (DCs) that are similar in functionality except that DC_1 warns the driver when it is backing towards an object behind the car while DC_2 does not. What is the structural basis of this warning function? Nearly all structural elements in DC_1 have their "homologues" in DC_2 except for the four sensors on the rear bumper of DC_1. This would lead us to quickly hypothesize that the four sensors are associated with the warning function, which turns out to be true. Now if we replace DC_2 with a baby stroller, then the comparison will be quite difficult because a stroller and a DC differ structurally in numerous ways and any structural difference could be responsible for the warning function. We may mistakenly hypothesize that the rear lights or the rear window defroster in DC_1, which are all missing in the stroller, may be responsible for the warning function. To test the hypotheses, we would destroy the rear lights, the rear window defroster, etc., one by one, but will get nothing but negative results. What could be even worse is that, when destroying the rear lights, we accidentally destroy a part of the electric system in such a way that the warning function is lost, which would mislead us to conclude that the rear lights are indeed part of the structural basis responsible for the warning function—an "experimentally substantiated" yet wrong conclusion. A claim that what is true in *E. coli* is also true in the elephant is equivalent to a claim that what is true in a stroller is also true in a DC. It will take comparative genomics out of its proper conceptual framework in evolutionary biology and render it inefficient to address biological questions.

Let's take a biologically more relevant example involving *Shigella flexneri* and *E. coli* (Sansonetti et al. 1982a, b). *Shigella* strains cause shigellosis, whereas strains of *Escherichia coli* are generally avirulent. What is responsible for the difference? Nuclear genomes are similar between *Shigella* and *E. coli*, which led scientists to focus on a plasmid that is present in the pathogenic *Shigella* strains but absent in the avirulent *E. coli* strains. The pathogenic *Shigella* strains become avirulent when the plasmid is taken away, and originally avirulent strains of *E. coli* gains virulence after acquiring the plasmid. This led quickly to the conclusion that the plasmid is largely responsible for shigellosis. Had one compared between *S. flexneri* and *Saccharomyces cerevisiae,* one would need to hypothesize that any one of the thousands of genes in *S. cerevisiae* not shared by *S. flexneri* could be a causal factor. Filtering through these thousands of possibilities would take forever even if we do not consider gene combinations as causal factors.

In this chapter I will detail a few typical comparative genomic studies so that we can develop an intuitive appreciation of what is hidden in the box labelled "comparative genomics". These studies involve biological problems that can be addressed by comparing two genomics as well as problems that would require more than two genomes to reach a solution. The similarities among these studies are summarized at the end to highlight essential elements in a comparative genomics study.

Genomic Comparison Between *Helicobacter pylori* and its Relatives

Problems and Hypotheses

Helicobacter pylori is a human pathogen causing gastric and duodenal ulcers and gastric cancer (Hamajima et al. 2004; Hunt 2004; Menaker et al. 2004; Siavoshi et al. 2004). It is an acid-resistant neutralophile (Bauerfeind et al. 1997; Rektorschek et al. 2000; Sachs et al. 1996; Scott et al. 2002) capable of surviving for at least 3 h at pH $= 1$ with urea (Stingl et al. 2001) and maintaining a nearly neutral cytoplasmic pH between pH 3.0 and 7.0 (Matin et al. 1996; Scott et al. 2002). In the presence of urea, *H. pylori* can accomplish its cytoplasmic pH homeostasis down to an external pH of 1.2 (Stingl et al. 2002b). These properties allow it to survive and reproduce in the human stomach where the gastric fluid has a pH averaging about 1.4 over a 24-h period (Sachs et al. 2003).

The buffering action of the gastric epithelium and limited acid diffusion through the gastric mucus were previously thought to protect the bacterium against stomach acidity, but both empirical studies (Allen et al. 1993) and theoretical modeling (Engel et al. 1984) have suggested that the protection is rather limited (Matin et al. 1996; Sachs 2003 #14944). Recently it has also been shown that mucus does not hinder proton diffusion and a trans-mucus pH gradient is abolished when the luminal pH drops to <2.5 (Baumgartner and Montrose 2004). It is therefore necessary for *H. pylori* to have acid-resisting mechanisms to colonize the gastric mucosa successfully (Sachs et al. 2003).

H. pylori has evolved two mechanisms protecting itself against the acidic environment in the mammalian stomach. The first, schematically illustrated in Fig. 1.1, involves the urease gene cluster *ureABIEFGH*. The constitutively expressed cytoplasmic urease consists of four heterodimer each with two subunits coded by *ureA* and *ureB*, respectively. It catalyzes urea to generate $2NH_3 + CO_2$ to buffer against the H^+ influx into either the periplasm or the cytoplasm (Mobley et al. 1991; Rektorschek et al. 2000; Sachs et al. 2003; Stingl et al. 2002a) and to facilitate the extrusion of H^+ from the cytoplasm in the form of NH_4^+

(Stingl et al. 2002a). However, urease is an apoenzyme requiring a nickel to be active. The *ureEFGH* gene cluster, whose expression is acid-induced, codes for nickel-sequestrating proteins that insert nickel into the urease, leading to increased and sustained urease activity (Sachs et al. 2003; Wen et al. 2003; Williams et al. 1996).

The urease, once activated, naturally needs a constant supply of urea as its substrate, and the cell has two sources of urea supply, one intrinsic and one extrinsic (Fig. 1.1). The extrinsic source refers to urea present in saliva and stomach fluid. The exposure of *H. pylori* to gastric acid results in a large increase in urea influx into the cell due to the pH-gating of the urea channel protein *UreI* (Bury-Mone et al. 2001; Weeks et al. 2000). The intrinsic source comes from efficient conversion of arginine to urea in the cytoplasm by the highly expressed arginase in *H. pylori* (Mendz and Hazell 1996). For this reason, arginine is underused, but lysine is overused, in *H. pylori* proteins (Xia and Palidwor 2005).

The second acid-resistant mechanism in *H. pylori* is the restriction of acute proton entry across its membranes by having a high frequency of positively charged amino acids and consequent high pI (isoelectric point) values in the inner and outer membrane proteins (Sachs et al. 2003; Scott et al. 1998; Valenzuela et al. 2003). This is supported by recent discovery of a basic proteome (Tomb et al. 1997), a set of basic membrane proteins (Baik et al. 2004) in *H. pylori*, and an extensive genomic analysis (Xia and Palidwor 2005) testing the adaptation, preadaptation and exaptation hypotheses concerning the overuse of lysine residues in *H. pylori* proteins. The mechanism gained functional importance after the discovery that urease-negative *H. pylori* can colonize the acidic gastric environment and cause gastric ulcers in Mongolian gerbils (Mine et al. 2005).

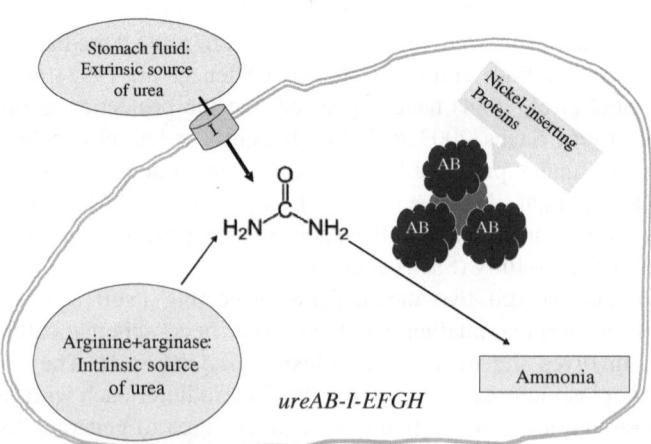

Fig. 1.1 Schematic illustration of the acid-resistance mechanisms in *H. pylori* mediated by genes in the urease gene cluster *ureAB-I-EFGH*

Given that *H. pylori* has many Lys-rich proteins with high pI values relative to other bacterial species that do not live in acidic environment, one is naturally tempted to conclude that the high pI values in the *H. pylori* proteins represent an adaptation to the acidic environment. However, there are at least four possible hypotheses for the origin of the basic proteome in *H. pylori* (Xia 2007a, Chap. 10).

The first hypothesis states that *H. pylori* would benefit from positively charged proteins (especially membrane proteins) to alleviate the influx of H^+ into cytoplasm. This hypothesis is known as the acid-adaptation hypothesis (Xia and Palidwor 2005), i.e., *H. pylori* acquired its high-pI proteins as an adaptation in response to selection imposed by the acidic environment.

The second hypothesis argues that parasitic bacterial genomes typically evolve towards AT-richness because spontaneous mutations are generally AT-biased according to comparisons between pseudogenes and their functional counterparts (Gojobori et al. 1982; Li 1983; Li et al. 1981) and the discovery of the prevalence of spontaneous C $\rightarrow$ T/U deamination (Frederico et al. 1990, 1993; Lindahl 1993). All known parasitic bacterial genomes are AT-rich. *H. pylori* has a relatively AT-rich genome, e.g., the genomic GC% of *H. pylori* 26695 is only 38 %, in contrast to the genomic GC% of 50 % in *E. coli* substr DH10B. The AT-richness would lead to an increase in A-rich codons such as the lysine codon AAA and AAG and a consequent increase in lysine usage and protein pI. Because *H. pylori* and its sibling species are all parasites, their most recent common ancestor might have already practiced parasitism, acquired AT-richness and increased frequency of lysine codons before it became a parasite in the mammalian stomach. Therefore, an overrepresentation of lysine residues in its proteins, if beneficial for acid-resistance, would represent an exaptation, i.e., the process in which an originally neutral trait has subsequently acquired a beneficial function. A well known example of exaptation is the brain-specific RNA gene BC200 resulting from the exaptation of a presumably neutral SINE repeat (Smit 1999).

The third hypothesis states that nucleotide C is rare in eukaryotic cells and a eukaryotic parasite should therefore minimize the usage of C as a building block for its RNA and DNA. CTP concentration is much lower than the other three nucleotides chick fibroblast cells (Colby and Edlin 1970) and in mouse 3T3 cells (Weber and Edlin 1971), suggesting the generality of C limitation. Consistent with the suggestion, the protozoan parasite, *Trypanosoma brucei*, maintains its *de novo* synthesis pathway for CTP and inhibiting its CTP synthetase effectively eradicates the parasite population in the host (Hofer et al. 2001). In contrast, the parasite does not have *de novo* synthesis pathways for purines, suggesting that the parasite can obtain the purines by its salvage pathway. This suggests that little CTP can be salvaged from the host. The relevance of these observations is highlighted by the fact that *H. pylori* maintains an active biosynthesis pathway, and a much less active salvage pathway, for pyrimidine nucleotides (Mendz et al. 1994). Thus, it might be evolutionarily beneficial for a mammalian parasite or symbiont to minimize the use of CTP in its DNA in building its genomes and in transcription (Rocha and Danchin 2002; Xia 1996).

Minimizing C in an organism with a DNA genome has the necessary consequence of reduced G, with a consequent increase in A and T. This will also contribute to increase AT and increased lysine codon. Thus, lysine overuse represents
a secondary consequence of an adaptation to a C-rare environment, but it predis-
posed the organism to tolerate an acidic environment. Such a mechanism is called
preadaptation, i.e., a trait originally selected for one function but that subsequently
gained a different function beneficial to the carrier of the trait. An often cited
example of preadaptation is the rudimentary feather that presumably has been
selected for thermoregulation in nonavian dinosaurs but preadapted their carriers
to subsequent evolution of flight.

The fourth hypothesis is more complicated. A protein in a solution with a pH
equal to the protein pI is not charged. If highly expressed proteins happen to have
their pI equal to the cytoplasmic pH, then there is no electrostatic repulsion among
these proteins when they are mass-produced. Such proteins will have low solubil-
ity and tend to aggregate and precipitate, which is often harmful to the cell. The
"amyloid precursor protein" causing Alzheimer disease and the prion protein causing the mad cow disease are examples of the undesirable protein aggregation and
precipitation. Take *E. coli* for example. Its intestinal environment has pH close to
9 and it can maintain its optimal growth at external pH as high as 8.8 (Zilberstein
et al. 1980, 1982). Its intracellular pH is regulated in the range of 7.4–7.8 at external
pH range of 5.5–9 (Slonczewski et al. 1981). Thus, in its intestinal environment, its
internal pH should be around 7.8 and we should expect *E. coli* to avoid having proteins with their pI values around 7.8. This is true (Fig. 1.2). Avoiding proteins with
pI equal to intracellular pH appears to be universal among unicellular organisms.

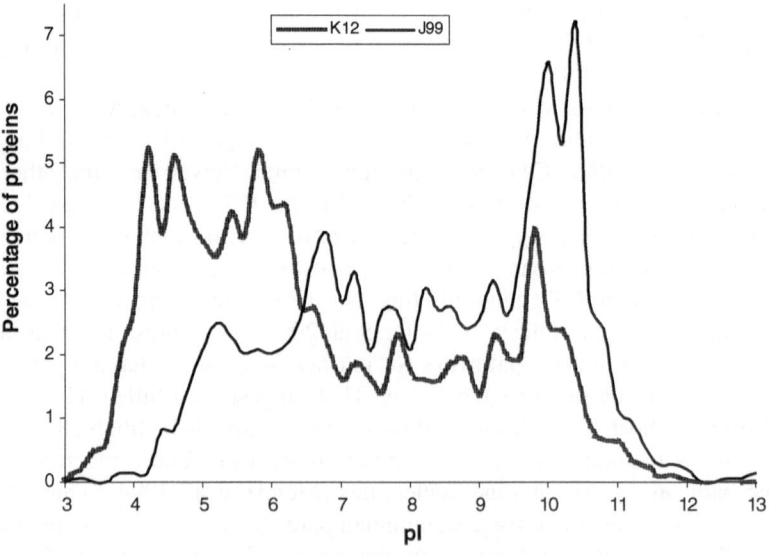

Fig. 1.2 Genomic pI profiling for *E. coli* K12 and *H. pylori* J99

Given the avoidance of proteins with pI equal to intracellular pH, we would expect mass-produced proteins in the gastric *H. pylori*, whose intracellular pH is around 5, to avoid having pI ≈ 5. This prediction is substantiated (Fig. 1.2). The pronounced peak of proteins with pI in the range of 4–6 in *E. coli* is missing in *H. pylori*. Instead, proteins with pI in the range of 10–11 are over-represented in *H. pylori* (Fig. 1.2)

One might ask why *H. pylori* proteins cannot lower their pI to the range of 0–3 to avoid precipitation. This would be practically difficult because the proteins would require an excessively large number of GAN to code for Glu and Asp. It is extremely rare to have proteins with a pI smaller than 3.

Testing the Hypotheses by Comparative Genomics

The first three hypotheses have been tested before and the second and third hypotheses were found to be inconsistent with the empirical data (Xia and Palidwor 2005). Here we illustrate how to discriminate between the first and the last hypothesis, i.e., whether the increase in protein pI is for alleviating the influx of protons, referred hereafter as AAH (acidity adaptation hypothesis), or for avoiding protein precipitation, referred to hereafter as precipitation avoidance hypothesis (PAH).

The two hypotheses have different predictions. AAH predicts that it is those membrane proteins that tend to gain a higher pI. In contrast, PAH predicts that the overrepresentation of the high-pI proteins in *H. pylori* is due to the necessity of mass-produced proteins to have their pI shifting away from the cytoplasmic pH to avoid protein precipitation. Specifically, the shifting of the pI distribution to the right in Fig. 1.2 is due to mass-produced proteins increasing their pI to shift their pI away from the cytoplasmic pH.

To test the AAH prediction, one needs to separate proteins into membrane proteins and cytoplasmic proteins. The main difficulty is that membrane proteins are difficult to separate and identify and only 34 membrane proteins have been identified in *H. pylori* (Baik et al. 2004). These proteins do exhibit a significantly higher pI than the rest of the *H. pylori* proteins (Xia and Palidwor 2005). Furthermore, one can use an excellent bioinformatic tool, pSort (Gardy et al. 2003; Nakai and Horton 1999), for protein cellular localization. Those proteins identified to be localized in cytoplasmic membrane, outer membrane and periplasmic space all have their mean pI values highly significantly higher than those localized in cytoplasm.

Are these results in favor of AAH? Not necessarily. Although AAH predicts that membrane proteins with a high pI would contribute to a positively charged shell alleviating the influx of protons into the cell, the result cannot be claimed to support, or even be consistent with, AAH. The reason is that membrane proteins in general have higher pI than cytoplasmic proteins, even for bacterial species that do not live in an acidic environment. What is important is to find bacterial species that are phylogenetically closely related to *H. pylori*, but do not exhibit acid resistance. Such species could be *H. hepaticus* or *Campylobacter* species, and are generally referred to as phylogenetic controls (because they and *H. pylori* were identical when we trace them back in time to their common ancestor). If we can find such pairs of sister species, with one living in acidic environment and the other not, and if we consistently find the former to have significantly elevated pI in their membrane proteins than the latter, then we can claim that the result supports, or at least is consistent with, the prediction of AAH. What is exciting about comparative genomics today is that, once we are equipped with the conceptual framework above, it takes only a few hours to complete the analysis by using publicly available genomic databases and software packages such as DAMBE (Xia 2001; Xia and Xie 2001). The empirical result, as you can verify by yourself, is consistent with the prediction of AAH. Both *H. pylori* and *H. hepaticus* have their membrane proteins with significantly high pI than cytoplasmic proteins, but the difference is much greater in *H. pylori* than in *H. hepaticus.*

Testing the prediction of PAH (i.e., mass-produced *H. pylori* proteins should evolve to have increased pI values away from cytoplasmic pH around 5) seems straightforward at first. We need to obtain pI and protein expression for each gene. Although we do not have reliable protein expression data in *H. pylori* at the moment, the difficulty can somewhat overcome by using indices of codon usage bias as a proxy of gene expression (Xia 1998a, 2007b, 2008). Similarly, although we do not have experimentally determined pI for each protein, theoretically derived pI based on amino acid composition (Xia 2007a, pp. 207–212) represents a good approximation. Now suppose we have protein pI and protein expression (designated by E). It seems that the prediction of PAH can be reduced to a statement that pI and E are positively correlated because high-E proteins should increase their pI away from the cytoplasmic pH. Is this inference correct? Now suppose you found that pI and E are indeed positively correlated, will you conclude that PAH is supported? Alternatively, if you found pI and E are negatively correlated, will you reject PAH?

It turns out that you cannot say much about PAH based on the correlation between pI and E. A positive correlation is expected if the data include many highly expressed DNA-binding or RNA-binding proteins because these proteins all tend to have a DNA/RNA-binding domain which is rich in positively charged amino acids (Recall that the backbone of RNA and DNA are negatively charged and a positively charged protein domain facilitates the binding to RNA and DNA). This would result in a positive correlation between pI and E which has nothing to do with PAH.

You may also get a negative correlation between pI and E for the following reason. Differences in pI among proteins mainly depend on the relative number of the strongly acidic amino acid residues such as Asp, and Glu and the strongly basic amino acid residues such as Arg, Lys, and His. The positively charged amino acids, however, are generally more energetically expensive to make in bacterial species (Akashi and Gojobori 2002). For example, the total high-energy ~P required to make Asp and Glu are 12.7 and 15.3, respectively, which are quite close to the cost of making the smallest amino acids Gly and Ala. In contrast, the energetic costs for making His, Lys and Arg are 38.3, 30.3 and 27.3, respectively. Highly expressed proteins tend to use cheap amino acids and avoid the expensive Arg, Lys and His in almost all bacterial species, resulting in highly expressed proteins (except for those ribosomal proteins) to have a low pI and a consequent negative correlation between pI and E. Thus, a negative correlation between pI and E again could have nothing to do with PAH.

Thus, to properly test the prediction of PAH, comparative genomics involving sister species (e.g., between *H. pylori* and *H. hepaticus*) is again necessary. Suppose we found 500 *H. hepaticus* proteins that have pI around 5 and are homologous to those in *H. pylori*. Also suppose that, among the 500 proteins, 200 of them are highly expressed and 200 are lowly expressed. If the 200 highly expressed proteins in *H. pylori* have all shifted their pI away from the cytoplasmic pH of about 5, whereas the 200 lowly expressed proteins have their pI hardly changed relative to their *H. hepaticus* homologues, then we can claim that result is consistent with PAH. Of course, this represents only one of possible ways to test the prediction from PAH.

Genomic Comparison Between HIV-1 and HTLV-1

Because viruses use the host translational machinery to translate their own mRNA, their codon usage is under selective pressure to adapt to the host tRNA pool (Sharp and Li 1987). In RNA viruses in general and Human Immunodeficiency Virus 1 (HIV-1) in particular, adaptation to the host is poor despite this selection (Bahir et al. 2009; van Weringh et al. 2011), in contrast to the codon-anticodon adaptation documented in bacterial genomes (Gouy and Gautier 1982; Ikemura 1981a, 1992; Xia 1998a) as well as in mitochondrial genomes in vertebrates (Xia 2005; Xia et al. 2007) and fungi (Carullo and Xia 2008; Xia 2008). For example, according to a recent compilation of tRNAs in human genome (Chan and Lowe 2009), the AUC codon can be translated by 17 tRNA[Ile] species, i.e., 14 tRNA[Ile/IAU] and 3 tRNA[Ile/GAU], AUU can be translated by 14 tRNA[Ile/IAU] species, whereas AUA can be translated by only 5 tRNA[Ile/UAU] species. In agreement with this, human genes code Ile mostly by AUC and least by AUA. In contrast, HIV-1 genes code Ile mostly by AUA and least by AUC (Haas et al. 1996; Nakamura et al. 2000). The poor codon adaptation of HIV-1 reduces the translation efficiency of HIV-1 genes. Modifying HIV-1 codon usage according to host codon usage has been shown to increase the production of viral proteins (Haas et al. 1996; Ngumbela et al. 2008).

The A-biased mutation hypothesis has been proposed to explain the poor concordance between HIV-1 and host codon usage (Jenkins and Holmes 2003). The A-bias is mediated by the error prone reverse transcriptase (Martinez et al. 1994; Vartanian et al. 2002) and the human APOBEC3 protein (Yu et al. 2004). The frequency of A can reach up to 40 % in some HIV-1 genomes (Vartanian et al. 2002), resulting in a preponderance of A-ending codons which are typically rarely used in the host genes (Kypr and Mrazek 1987; Sharp 1986). While there have been claims that the A-richness in a parasitic or symbiotic genome may confer some selective advantage (Keating et al. 2009; Xia 1996), further empirical substantiation is required. In short, although avoiding A-ending codons will lead to better codon-anticodon adaptation, strongly A-biased mutations lead to an over-representation of A-ending codons in HIV-1 genes, disrupting codon-anticodon adaptation.

How can we test this mutation hypothesis? If we can find pairs of sister species that differ much in mutation rate, then we can test the hypothesis by checking if the species with higher mutation rate tend to have poorer codon-anticodon adaptation than its sister species with lower mutation rate. HTLV-1 could serve as a sister species for the HIV/SIV lineage. Both HTLV-1 and HIV-1 are retroviruses with RNA genomes and both infect the same type of host cell, i.e., human CD4 + T cells (Rimsky et al. 1988). The two viruses are therefore subject to the same selective pressures on codon usage by the host tRNA pool. However, HTLV-1 is exceptional in that it does not have a strong A-biased mutation spectrum (Van Dooren et al. 2004; van Hemert and Berkhout 1995). HTLV-1 relies for the most part on the host polymerase to replicate through clonal expansion of infected cells rather than undergoing iterative replication cycles like HIV-1 (Strebel 2005). The substitution rate of HTLV-1 is consequently lower, about 5.2×10^{-6} substitutions/site/year (Hanada et al. 2004; Van Dooren et al. 2004), in contrast to that of HIV-1 at 2.5×10^{-3} substitutions/site/year (Hanada et al. 2004). Codon-anticodon adaptation is less likely to be disrupted by mutation in HTLV-1 than in HIV-1. Thus we predict that HTLV-1 coding sequences should exhibit better codon-anticodon adaptation.

Codon-anticodon adaptation can be measured by the correlation in RSCU (Sharp and Li 1987) between the host and the parasite. RSCU is a normalized index of codon usage (Sharp and Li 1987). It has a value of zero for unused synonymous codons, a value of one for equally used synonymous codons and a maximum of n, where n is the number of synonymous codons in the codon family. Thus, the prediction of the mutation hypothesis is that the correlation in RSCU between human and HTLV-1 genes should be greater than that between human and HIV-1 genes.

The correlation in RSCU between human and HIV-1 genes is poor (Pearson $r = -0.1470$, $p = 0.2665$; Spearman $r = 0.1829$, $p = 0.1657$). In contrast, the positive correlation in RSCU between HTLV-1 and human genes is highly significant (Pearson $r = 0.4982$, $p < 0.0001$, Spearman $r = 0.4688$, $p = 0.0002$). Such results are consistent with the mutation hypothesis.

The real scenario of codon-anticodon adaptation in HIV-1 is much more complicated, of course. In particular, the early gene and late genes in HIV-1 may be translated in different tRNA pools and subject to different selection for codon-anticodon adaptation (van Weringh et al. 2011). HIV-1 has recently been shown to package non-lysyl tRNAs in addition to the tRNA^{Lys} needed for priming reverse-transcription and integration of the HIV-1 genome. In particular, tRNAs decoding A-ending codons, required for the expression of HIV's A-rich genome, are highly enriched. Because the affinity of Gag-Pol for all tRNAs is non-specific, HIV packaging is most likely passive and reflects the tRNA pool at the time of viral particle formation. Codon usage of HIV-1 early genes is similar to that of highly expressed host genes, but codon usage of HIV-1 late genes were better adapted to the selectively enriched tRNA pool, suggesting that alterations in the tRNA pool are induced late in viral infection. If HIV-1 genes are adapting to an altered tRNA pool, codon adaptation of HIV-1 may be better than previously thought (van Weringh et al. 2011).

Genomic Comparison Among *Mycoplasma* Species

CpG deficiency has been documented in a large number of genomes covering a wide taxonomic distribution (Cardon et al. 1994; Josse et al. 1961; Karlin and Burge 1995; Karlin and Mrazek 1996; Nussinov 1984). DNA methylation is one of the many hypotheses proposed to explain differential CpG deficiency in different genomes (Bestor and Coxon 1993; Rideout et al. 1990; Sved and Bird 1990). It features a plausible mechanism as follows. Methyltransferases in many species, especially those in vertebrates, appear to methylate specifically the cytosine in CpG dinucleotides, and the methylated cytosine is prone to mutate to thymine by spontaneous deamination (Frederico et al. 1990; Lindahl 1993). This implies that CpG would gradually decay into TpG and CpA, leading to CpG deficiency and reduced genomic GC%. Different genomes may differ in CpG deficiency because they differ in methylation activities, with genomes having high methylation activities exhibiting stronger CpG deficiency than genomes with little or no methylation activity.

In spite of its plausibility, the methylation-deamination hypothesis has several major empirical difficulties (Cardon et al. 1994), especially in recent years with genome-based analysis (Goto et al. 2000). For example, *Mycoplasma genitalium* does not seem to have any methyltransferase and exhibits no methylation activity, yet its genome shows a severe CpG deficiency. Therefore, the CpG deficiency in *M. genitalium*, according to the critics of the methylation-deamination hypothesis, must be due to factors other than DNA methylation.

A related species, *M. pneumoniae*, also devoid of any DNA methyltransferase, has a genome that is not deficient in CpG. Given the difference in CpG deficiency between the two Mycoplasma species, the methylation hypothesis would have predicted that the *M. genitalium* genome is more methylated than the *M. pneumoniae* genome, which is not true as neither has a methyltransferase. Thus, the methylation hypothesis does not seem to have any explanatory power to account for the variation in CpG deficiency, at least in the Mycoplasma species.

These criticisms are derived from phylogeny-free reasoning. When phylogeny-based comparisons are made, the Mycoplasma genomes become quite consistent with the methylation hypothesis (Xia 2003). First, several lines of evidence suggest that the common ancestor of *M. genitalium* and

M. pneumoniae have methyltransferases methylating C in CpG dinucleotides, and should have evolved strong CpG deficiency and low genomic GC% as a result of the specific DNA methylation. Methylated m^5C exists in the DNA of a close relative, *Mycoplasma hyorhinis* (Razin and Razin 1980), suggesting the existence of methyltransferases in *M. hyorhinis*. Methyltransferases are also present in *Mycoplasma pulmonis* which contains at least four CpG-specific methyltransferase genes (Chambaud et al. 2001). Methyltransferases are also found in all surveyed species of a related genus, Spiroplasma (Nur et al. 1985). These lines of evidence suggest that methyltransferases are present in the ancestors of *M. genitalium* and *M. pneumoniae*.

Second, the methyltransferase-encoding *M. pulmonis* genome is even more defi-cient in CpG and lower in genomic GC% than *M. genitalium* or *M. pneumoniae*, consistent with the methylation hypothesis (Fig. 1.3). It is now easy to understand that, after the loss of methyltransferase in the ancestor of *M. genitalium* and *M. pneumoniae* (Fig. 1.3), both genomes would begin to accumulate CpG dinucleo-tides and increase their genomic GC%. However, the evolutionary rate is much faster in *M. pneumoniae* than in *M. genitanlium* based on the comparison of a large number of protein-coding genes (Xia 2003). So *M. pneumoniae* regained CpG dinucleotide and genomic GC% much faster than *M. genitalium*. In short, the Mycoplasma data that originally seem to contradict the methylation hypothesis actually provide strong support for the methylation hypothesis when phylogeny-based genomic comparisons are made.

One might note that *Ureaplasma urealyticum* in Fig. 1.3 is not deficient in CpG because its $P_{CpG}/(P_C P_G)$ ratio is close to 1, yet its genomic GC% is the lowest. Has its low genomic GC% resulted from CpG-specific DNA methylation? If yes, then why doesn't the genome exhibit CpG deficiency? It turns out that *U. urealyti-cum* has C-specific, but not CpG-specific, methyltransferase, i.e., the genome of *U. urealyticum* is therefore expected to have low CG % (because of the methyla-tion-mediated C → T mutation) but not a low $P_{CpG}/(P_C P_G)$ ratio. The methyltrans-ferase gene from *U. urealyticum* is not homologous to that from *M. pulmonis*.

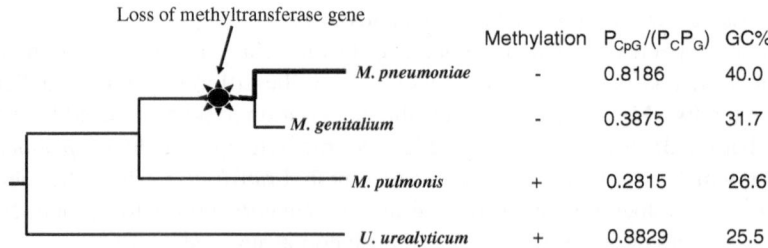

Fig. 1.3 Phylogenetic tree of *Mycoplasma pneumoniae*, *M. genitalium*, and their relatives, together with the presence (+) or absence (−) of CpG-specific methylation, PCpG/(PCPG) as a measure of CpG deficiency, and genomic GC%. *M. pneumoniae* evolves faster and has a longer branch than *M. genitalium*

We have seen how phylogeny can help us in evolutionary inference, and most comparative genomic studies represent phylogeny-based inference. It is appropriate here to introduce a few phylogeny-related terms. Most published phylogenies are build from molecular sequence data, i.e., multiple alignment of homologous sequences. Sequence similarity can arise in two ways, one from convergence (i.e., similarity gained from independent evolution), and the other from coancestry. Coancestral sequences are homologous, and can be divided into orthologous and paralogous sequences. Two or more duplicated genes within one genome represent a special form of homology and are termed paralogous genes. Two or more homologous genes that are related by inheritance are orthologous. Genes acquired through horizontal gene transfer are neither orthologous nor paralogous. Species phylogeny ideally should be built only from orthologous genes.

Genomic Comparison to Characterize Changes in tRNA and Codon-Anticodon Adaptation

Ever since the empirical documentation of the correlation between codon usage and tRNA abundance (Ikemura 1981a, b, 1982, 1992), studies on codon-anticodon adaptation have progressed in theoretical elaboration (Bulmer 1987, 1991; Higgs and Ran 2008; Jia and Higgs 2008; Palidwor et al. 2010; Xia 1998a, 2008), in critical tests of alternative theoretical predictions (Carullo and Xia 2008; Plotkin and Kudla 2010; Plotkin et al. 2004; van Weringh et al. 2011; Xia 1996, 2005) and in formulation and improvement of various codon usage indices to characterize codon usage bias (Sharp and Li 1987; Wright 1990; Xia 2007b). Here I present two examples in which a gain/loss of a tRNA gene or a change in genetic code lead to significant changes in codon usage.

The Met Codon Family

An evolutionary change in tRNA composition or relative abundance is expected to alter codon-anticodon adaptation. This is not controversial theoretically. However, how fast can an alternation in tRNA lead to consequent changes in codon-anticodon adaptation? Can the cause-effect relationship be demonstrated with empirical data? Changes in tRNA^{Met} genes (where Met is the amino acid carried by the tRNA) in animal mitochondrial DNA (mtDNA) paved the way for such a demonstration (Xia 2012b).

In MtDNA of most animal species, Met is coded by AUA and AUG codons. In some animal species, e.g., vertebrates, these two codons are translated by a single tRNA$^{Met/CAU}$ species (where CAU is the anticodon in the 5' to 3' orientation) with a modified C (i.e., f^{5}C) at the first anticodon position (Grosjean et al. 2010) to allow C/A pairing. In other animal species, e.g., tunicates, an additional tRNA$^{Met/UAU}$ gene is present in the mtDNA. One would expect that, when tRNA$^{Met/UAU}$ is absent, Met should be preferably coded by AUG with a reduced AUA usage. The gain of tRNA$^{Met/UAU}$ would favor more Met to be coded by AUA. Can such a prediction be empirically substantiated?

MtDNA in bivalve species have two tRNA^{Met} genes. In some bivalve species (e.g., *Acanthocardia tuberculata, Crassostrea gigas, C. virginica, Hiatella arctica, Placopecten magellanicus,* and *Venerupis philippinarum*), both tRNA^{Met} genes have a CAU anticodon forming Watson–Crick base pair with codon AUG. In some other bivalve species (e.g., *Mytilus edulis, Mytilus galloprovincialis,* and *Mytilus trossulus*), one tRNA^{Met} has a CAU anticodon and the other has a UAU anticodon forming Watson–Crick base pair with the AUA codon. One would predict that the latter should be more likely to code Met by AUA than the former, i.e., the proportion of AUA codon within the AUR codon family, designated P_{AUA}, should be greater in the latter with both a tRNA$^{Met/CAU}$ and a tRNA$^{Met/UAU}$ gene than in the former with a single tRNA$^{Met/CAU}$ gene in the mtDNA (Xia et al. 2007).

To test the prediction, I will use P_{UUA} (the proportion of UUA codon in the UUR codon family) as a reference control to test the prediction that, at the same P_{UUA} level, P_{AUA} in the three Mytilus mtDNA with both a tRNA$^{Met/CAU}$ and a tRNA$^{Met/UAU}$ gene is higher than that in the six bivalve species without a tRNA$^{Met/UAU}$ gene. This is supported by empirical evidence (ANCOVA test, $p = 0.0111$, Fig. 1.4a). Thus, the presence of tRNA$^{Met/UAU}$ increases AUA usage significantly.

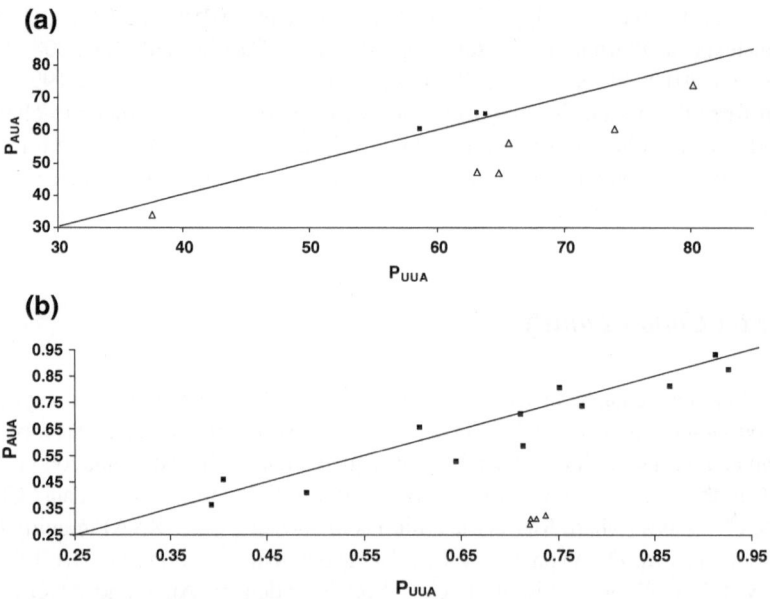

Fig. 1.4 Relationship between PAUA and PUUA, highlighting the observation that PAUA is greater when both a tRNAMet/CAU and a tRNAMet/UAU are present than when only tRNAMet/CAU is present in the mtDNA, for bivalve species (**a**) and chordate species (**b**). The filled squares are for mtDNA containing both tRNAMet/CAU and tRNAMet/UAU genes, and the open triangles are for mtDNA without a tRNAMet/UAU gene

A similar comparison can be performed between the urochordates (tunicates, with both tRNA$^{Met/CAU}$ and tRNA$^{Met/UAU}$ genes in their mtDNA) and cephalochordates (lancelets, with only a tRNA$^{Met/CAU}$ gene in their mtDNA). Figure 1.4b shows that P_{AUA} is much smaller in lancelets than in tunicates at the same P_{UUA} level. Thus, AUA usage is consistently increased by the gain of a tRNA$^{Met/UAU}$ gene (or consistently decreased by the loss of a tRNA$^{Met/UAU}$ gene) in animal mtDNA.

A gain of a tRNA$^{Met/UAU}$ gene is also associated with a surplus of AUG → AUA substitutions in animal mitochondrial coding sequences (results not shown). Similar associations can also be observed with other gain/loss of tRNA genes in animal mitochondrial. In contrast, a gain/loss of tRNA genes in plant mtDNA appears to have little effect on nucleotide substitutions or codon usage, presumably because such gain/loss events do not significantly alter the tRNA pool in plant cells where nuclear tRNAs are mass-imported into plant mitochondria.

UGA Codon, CGN Codon for Arg and the Expanded Wobble Hypothesis

The number of distinct tRNA species is invariably fewer than the number of sense codons, leading to the formulation of the original wobble hypothesis (Crick 1966). Figure 1.5 depicts the extended codon-anticodon base pairs as well as the subscripted numbering system used for codon-anticodon base pairs (Xia 2013). Note that the anticodon sites are denoted by Roman numerals and tho the codon sites by Arabic numerals (Fig. 1.5).

The wobble hypothesis explains why tRNA$^{Ile/IAU}$, where I in IAU is inosine derived from A, is able to translate all three Ile codons (AUC, AUU and AUA), why a tRNA with a G_I can translate Y-ending codons (where Y stands for C or U), and why a tRNA with a U_I can translate R-ending codons (where R stands for A or G). The hypothesis also explains the lack of A_I in tRNA genes for decoding 2-fold Y-ending codon family because such a tRNA, when its A_I is modified to I_I, would mis-read the near cognate R-ending codons. One might note that all base-pairs involve a purine and a pyrimidine except for the I/A pair which is a bulky purine-purine pair that may lead to inefficient translation (Curran 1995).

Wobble pairing reduces the number of tRNAs needed for translation and simplifies the translation machinery. Few organisms can afford the luxury of having different gene products doing the same task. As an example of parsimonious tRNA usage, the Y-ending codons, be they in 2-fold or 4-fold codon families, are decoded by tRNAs with either a I_I or a G_I, but never both. This rule is obeyed in all three kingdoms of life. Almost all 4-fold codon families in *Mycoplasma pulmonis* (including the Ser UCN codon family and Leu CUN codon family, where N is any nucleotide) are decoded by a single tRNA species with a U_I, except for the Thr ACN and Arg CGN codon families which are each decoded by two tRNA species, one with a U_I and other with a G_I. The most dramatic simplification of tRNome is observed in metazoan mitochondria, e.g., vertebrate mitochondrial genomes which

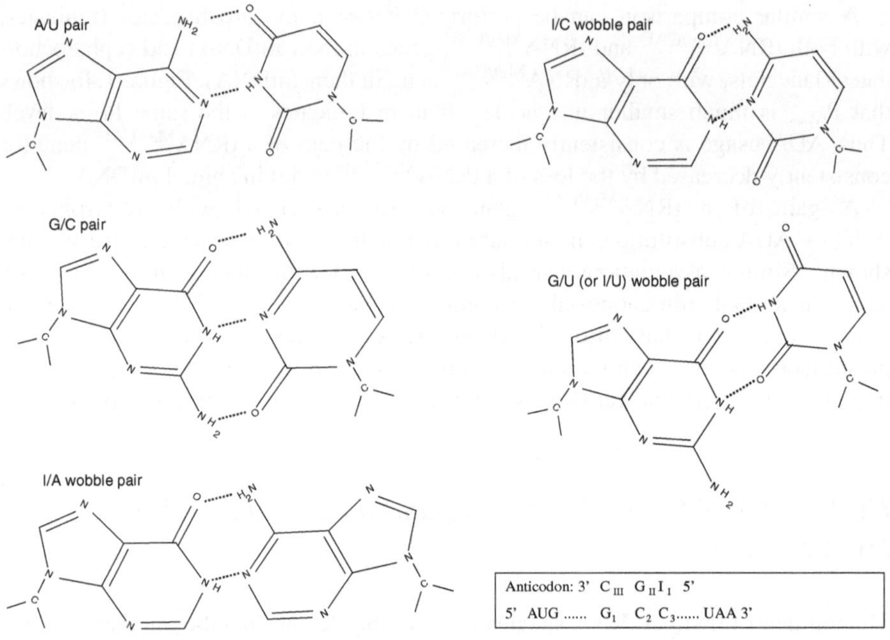

Fig. 1.5 Base pairs between nucleotides at the first anticodon site (which can have I, G, C, U but rarely A) and the third codon site. The inset shows the site numbering system of codon-anticodon base pairs, with codon sites subscripted with 1, 2 and 3 and anticodon sites subscripted with I, II, and III (corresponding to 34, 35 and 36 in the conventional system), so that base pairs are I_I/C_3, G_{II}/C_2, C_{III}/G_1. This numbering system is used because the 34th site of many tRNAs sequences is not the first anticodon site

contain only 22 tRNA genes, with each tRNA species decoding a codon family. Instead of separate initiation tRNA$^{iMet/CAU}$ and elongation tRNA$^{eMet/CAU}$ contained in all nuclear genomes, a single tRNA$^{Met/CAU}$, with a modified C_I, decodes both the initiation AUG codon and internal Met AUR codons. Each Y-ending codon family is decoded by a single tRNA species with a wobble G_I, and each R-ending codon family by a single tRNA with a wobble U_I which is modified to prevent its pairing with U or C. All 4-fold codon families are decoded by a tRNA with a wobble U_I which is not modified.

Recent comparative genomic studies on tRNA have led to the expanded wobble hypothesis (Carullo and Xia 2008; Xia 2013) which arose from the following observation. A tRNA species with a wobble U_I (where subscripted I indicates the first anticodon position that pairs with the third codon position) is almost always present among tRNA species decoding 4-fold codon families and 2-fold R-ending codon families, with most exceptions observed in the Arg CGN codon family. In the mitochondrial genomes of *Caenorhabditis elegans* (metazoan), *Marchantia polymorpha* (plant), *Pichia canadensis* (fungus), and *Saccharomyces cerevisiae* (fungus), there is no tRNA$^{Arg/UCG}$, and Arg CGN codon family is decoded by tRNA$^{Arg/ACG}$ (Xia 2005). The lack of tRNA$^{Arg/UCG}$ in the mitochondrial genome of these diverse taxa suggests that the lack is an ancestral state and that the presence of tRNA$^{Arg/UCG}$ in vertebrate mitochondria is a derived state. This is consistent with the observation that almost all eubacterial species, from which the mitochondrion was originally derived, lack tRNA$^{Arg/UCG}$ (Grosjean et al. 2010).

Why tRNA$^{Arg/UCG}$ is missing in the ancestral mitochondrial lineages and why did it appear in derived lineages such as vertebrate mitochondrial genomes? It is these questions that prompted the proposal of an expanded wobble hypothesis.

The expanded wobble hypothesis for the lack of tRNA$^{Arg/UCG}$ in bacterial and early mitochondrial lineages invokes wobble paring between the third anticodon site (X_{III}) and the first codon site (Y_1), conditional on a C_{II}/G_2 or G_{II}/C_2 with three hydrogen bonds. Thus, the anticodon UCG would wobble-pair with stop codon UGA through a wobble U_{III}/G_1 pair, and should therefore be strongly selected against because it would read through the stop codon (Carullo and Xia 2008). This not only explains the absence of tRNA$^{Arg/UCG}$ in bacterial and early mitochondrial lineages where UGA is used as a stop codon, but also why it appeared in derived mitochondrial lineages such as vertebrate mitochondrial genomes where UGA is no longer used as a stop codon. Wobble pairing involving N_{III}/N_1 represents a fundamental deviation from the original wobble hypothesis and requires further empirical validation.

Genomic Strand Asymmetry and Genome Replication

Most mutations occur during DNA replication, and different DNA replication mechanisms often leave distinct footprints in genomic strand asymmetric patterns because DNA polymerase for the leading and lagging strands differ in replication fidelity (Marin and Xia 2008; Xia 2012a). Strand asymmetry is typically measured by the GC skew (Lobry 1996; Marín and Xia 2008) defined as

$$S_G = \frac{P_G - P_C}{P_G + P_C} \tag{2.1}$$

A more general motif skew (Lopez et al. 1999) is defined as

$$S_m = \frac{N_m - N_{m_{rc}}}{N_m + N_{m_{rc}}} \tag{2.2}$$

where m is either a nucleotide (e.g., G or A) or a motif (e.g., ACG), m_{rc} is the reverse complement of m (m_{rc} = C if m = G, or m_{rc} = CGT if m = ACG), and N_x is the number of x (where x is either m or m_{rc}). GC skew and AT skew are special cases of S_m when m is equal to either G or A, respectively, i.e., GC Skew is S_G and AT skew is S_A. Strand asymmetry represents a primary feature of DNA genomes, and its study can lead to insight into different genome replication mechanisms. Strand asymmetry represents a primary feature of DNA genomes, and its study can lead to insight into different genome replication mechanisms. A typical S_G plot (Fig. 1.6a) allows one to infer the origin and termination of the replication fork.

Bacterial species from *Bacillus subtilis* to *Escherichia coli* share the strand asymmetric pattern in Fig. 1.6a, which is characteristic of the single-origin bi-directional DNA replication shared by eubacterial species, with the leading strand being GT-rich and lagging strand AC-rich. Interestingly, primitive forms of plants such as the liverwort *Marchantia polymorpha*, or primitive forms of metazoans such as the sponge *Oscarella lobularis,* have strand asymmetric patterns (Fig. 1.6b) that are indistinguishable from what is typically seen in bacterial genomes with a single origin of replication. This similarity in strand asymmetric patterns suggests similarity in replication mechanisms and may explain the extremely slow rate of evolution in primitive animal and plant mtDNA relative to mtDNA in higher meta-zoans. In other words, mitochondrial genomes in plants and primitive invertebrates may maintain the high-fidelity replication as in their bacterial ancestor.

The fast evolving vertebrate mtDNAs share the strand asymmetric pattern (Fig. 1.6c–d) consistent with the strand-displacement model of DNA replication (Bogenhagen and Clayton 2003; Brown et al. 2005; Clayton 1982, 2000; Shadel

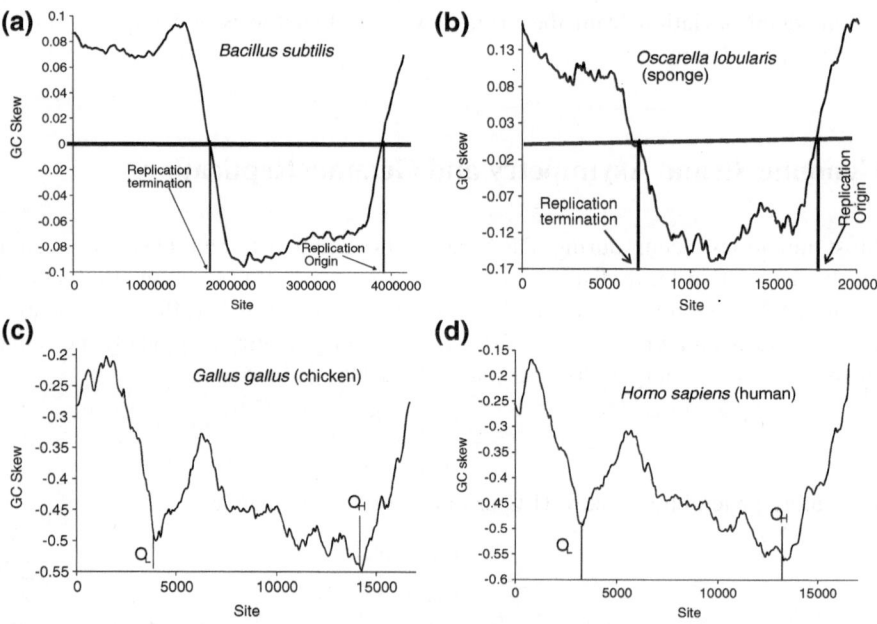

Fig. 1.6 Genomic strand asymmetric patterns characterized by GC skew values along a sliding window, with inferred replication origins. The *Bacillus subtilis* pattern (**a**) is shared among all eubacterial species known to have single-origin bi-directional replication. The sponge mtDNA, which evolves slower than the nuclear DNA, has the strand asymmetric pattern similar to its eubacterial ancestor (**b**). Vertebrate mtDNAs are replicated by the highly derived, but error-prone, two-origin strand-displacement replication, and evolve much faster than the nuclear DNA. Modified from Fig. 1, Fig. 9a, Fig. 9b and Fig. 10c in (Xia 2012a)

and Clayton 1997) which, although challenged recently by a new proposal of strand-coupled bidirectional replication (Yang et al. 2002; Yasukawa et al. 2005), is favored by current empirical evidence (Brown et al. 2005). According to this replication model, the L-strand is first used as a template to replicate the daughter H-strand, starting at the origin of replication O_H, while the parental H-strand was left single-stranded for an extended period because the complete replication of mtDNA takes nearly two hours (Clayton 1982, 2000; Shadel and Clayton 1997). After about 2/3 of the daughter H-strand has been synthesized and the second origin of replication (O_L) is exposed, the parental H-strand is used as a template to synthesize the daughter L-strand. Thus, different parts of the H-strands are in single-stranded form for different periods of time.

Single-stranded DNA binding proteins (SSB) protects single-stranded DNA from nucleolytic degradations. In *E. coli*, this works best with the presence of Rec-A. SSB from *E. coli* also reduces the C-U deamination rate in single-stranded DNA by 4–5 fold (Lough et al. 2001). However, it is not known if mtSSB also has the equivalent Rec-A partner or if it also protects single-stranded DNA from deamination in mitochondria.

Spontaneous deamination of both A and C (Lindahl 1993; Sancar and Sancar 1988) occurs frequently in human mtDNA (Tanaka and Ozawa 1994). Deamination of A leads to hypoxanthine that pairs with C, generating an A/T $\rightarrow$ G/C mutation. Deamination of C leads to U, generating C/G $\rightarrow$ U/A mutations. Among these two types of spontaneous deamination, the C $\rightarrow$ U mutation occurs more frequently than the A $\rightarrow$ G mutation (Lindahl 1993). In particular, the C $\rightarrow$ U mutation mediated by the spontaneous deamination occurs in single-stranded DNA more than 100 times as frequent as double-stranded DNA (Frederico et al. 1990). Note that these C $\rightarrow$ U sites will immediately be used as template to replicate the daughter L-strand, leading to a G $\rightarrow$ A mutation in the L-strand after one round of DNA duplication. Such mutation patterns are expected to leave their footprints on different parts of the H-strands left single-stranded for different periods of time.

While experimental evidence for the strand-displacement model is limited to mammalian species, the nearly identical pattern of strand asymmetry among vertebrate species suggests that the replication mechanism is most likely shared (Xia 2012a). The reduction in S_G correspond to the reduction of C in the H strand (and the associated G in the L strand), allowing us to infer the location of replication origins O_H and O_L (Fig. 1.6c–d). The GC skew values for vertebrate mtDNA are all negative, implying global asymmetry in addition to the local asymmetric patterns.

Strand asymmetry patterns provide an empirical test for inferred genome rearrangement by maximum parsimony. Much of the genome rearrangement in bacterial species may be attributed to inversion which leads to involved genes switching strands and experiencing different mutation spectrum. When two genomes or two genome segments with the same set of genes but differ in gene order, then one can compute the inversion distance which is the minimum number of inversions that can transform the gene order in one genome into that of

Table 1.1 Components of a comparative genomic study

Target genomes	Phylogenetic control	Genomic features	Biological problem involving genomic features
H. pylori	*H. hepaticus*	Protein pI, genomic GC%	Is protein pI increase in *H. pylori* driven by genomic GC% or by acid-adaptation?
HIV-1	HTLV-1	Codon adaptation, genomic mutation bias	Is poor codon-anticodon adaptation in HIV-1 caused by high mutation rate?
Mycoplasma species	Closely related species	CpG deficiency, methyltransferase, evolutionary rate	Is genomic CpG deficiencies driven by methylation-mediated mutation bias?
Bivalves, chordates	Closely related species	Codon usage, presence/absence of tRNA$^{Met/UAU}$	Does codon usage in met codon family evolve in response to the presence/absence of tRNA$^{Met/UAU}$?

the other (Kececioglu and Sankoff 1994, 1995). When the inversion event is rare, then this maximum parsimony approach is reasonable. However, it is important to keep in mind that the inferred inversion events constitute only a hypothesis that needs to be empirically tested. Because inversion events would leave its footprints in strand asymmetry patterns, we can test the hypothesis by checking whether the strand asymmetry pattern is consistent with the inferred inversion events.

In summary, a comparative genomic study contains four essential elements: (1) genomes with biologically interesting genotypic or phenotypic traits, (2) phylogenetic control, (3) genomic features, and (4) a solvable biological problem involving genomic features. These components are summarized in Table 1.1 for the four studies outlined in this chapter. Many comparative genomics studies focus on the gene order as a genomic feature to understand how various recombination mechanisms would lead to gene and exon reshuffling. Phylogenetic controls are particularly important for such genome rearrangement studies because one can reconstruct genome rearrangement events reliably only with very closely related genomes with few rearrangement events.

Chapter 2
Comparative Genomics and the Comparative Methods

Large-scale comparative genomics involves the type of data aimed to understand functional association among genes, between genes and phenotypes, and between genotype/phenotype and the environmental variables (Fig. 2.1). The most straightforward genetic variables (referred hereafter as G variables) are the presence/absence of genes, so the G_{ij} variables will be binary. Alternatively, A G variable could be a polymorphic site in a set of aligned HIV-1 polyproteins and take one of 20 alternative states. Changes in G_i may be associated with changes in G_j (e.g., a loss of a negative charge at site i may be compensated by a gain of a negative charge at a neighbouring site). Similarly, a change in G_i may have a phenotypic effect, i.e., G_i is associated with one or more of the phenotypic variables (hereafter referred to as P variables). For example, a change in G_i may results in drug resistance. Evolutionary biologists are also interested in whether certain changes in the G variables and P variables are in response to the environmental variables (hereafter referred to as E variables).

Depending on one's research objective, one may also allow the G variables to take continuous values. For example, with N orthologous protein-coding genes shared among all genomes, G_i can be the isoelectric point (pI) for protein i. It is known that pI of a number of enzymes co-evolve with the pI of their substrates, simply because an enzyme and its substrate typically should not be both positively charged or both negatively charged - they would push each other apart if they were. In the simplest case, we could have a single G variable representing pI of, say, laccase, and a single P variable representing the optimal pH for laccase. Such a setup would allow us to study the association between the G and P variables.

Dr. Tianjue Hu (a former postdoctoral fellow in my lab) carried out an interesting study relating pI of lignin-degrading laccases to the optimal pH of the enzyme. Different laccases have been isolated from different fungal species living in environments with different pH. Because lignin is relatively hydrophobic, a laccase needs to be hydrophobic as well, which implies that its pI should be close to its environmental pH (pH_e). However, pI computed from the original protein specified in the coding sequence (pI_o) is often much higher than pH_e, suggesting that

X. Xia, *Comparative Genomics*, SpringerBriefs in Genetics,
DOI: 10.1007/978-3-642-37146-2_2, © The Author(s) 2013

Fig. 2.1 Typical data sets for large-scale comparative genomics, with 8 species/genomes and the associated genetic variables (G), phenotypic variables (P) and environmental variables (E)

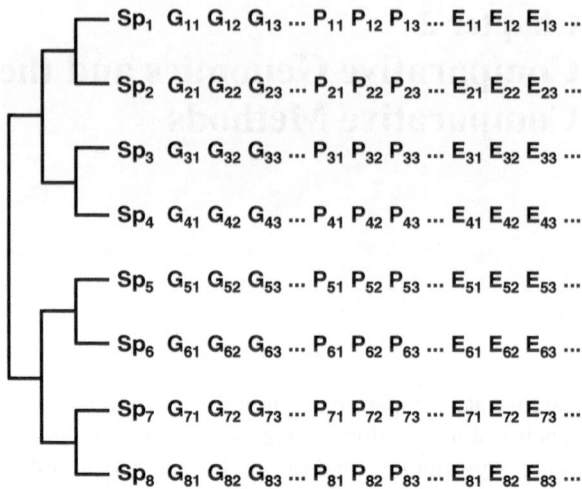

the nascent protein is not in its optimal state for digesting lignin and that there is selection on the organism to modify its laccase to reach a pI that is closer to pH_e. We may define an index of selection (I_s) as

$$I_s = pI_o - pH_e \tag{2.1}$$

We designate the pI of the modified mature protein as pI_m. The index of response to the selection by modifying laccase to reduce its pI can be defined as

$$I_r = pI_o - pI_m \tag{2.2}$$

We expect I_s and I_r to be positively correlated. The empirical evidence (unpublished) does support this expectation. Comparative genomics, with the data illustrated in Fig. 2.1, allows us to carry out millions of such studies simultaneously.

The G variables could also be the length of poly(A) tracts upstream of the translation initiation codon for a set of homologous genes shared among a set of eukaryotic genomes. Short pre-AUG poly(A) can bind to translation initiation factors and enhance translation initiation, whereas long pre-AUG poly(A) would bind to the poly(A)-binding protein and inhibit translation initiation (Xia et al. 2011). While the translation machinery is highly conserved in fungal species, the length and location of pre-AUG poly(A) of mRNAs often vary much among homologous genes, leading to associated changes in relative gene expression (which is a P variable). We can thus study the relationship between the variation in pre-AUG poly(A), which is a G variable, and gene expression. In large-scale comparative genomics, the most frequently used continuous variables will most likely be SNP sites and copy number of mRNAs. For example, Affymetrix Genome-Wide Human SNP Array 6.0 produces both the discrete nucleotide state of SNP sites and the copy number of probes representing genes.

In what follows, I will numerically illustrate the comparative methods (Barker and Pagel 2005; Felsenstein 1985; Harvey and Pagel 1991; Pagel 1994; Schluter et al. 1997) for characterizing the association between any two columns of data shown in Fig. 2.1. The simplest data set in Fig. 2.1 would have only two columns of data, and that is the type of data I will use to illustrate the comparative method for the continuous and discrete variables. According to the late population geneticist C. C. Li, it is not necessary to create a rainbow spanning the sky to demonstrate how a rainbow forms—a small one is convincing enough.

Note that N columns of data would imply $N*(N-1)/2$ pairwise associations, so large-scale comparative genomic studies almost always lead to multiple comparisons. So I will also illustrate the computation involved in controlling for false discovery rate which represents a key development in recent studies of statistical significance tests (Benjamini and Hochberg 1995; Benjamini and Yekutieli 2001).

One evolutionary process that has shaped bacterial genomes is the horizontal gene transfer. The phylogenetic incongruence test used to detect such horizontal gene transfer events will also be illustrated.

The Comparative Method for Continuous Characters

The two continuous variables that I will use here is genomic GC% and optimal growth temperature in bacterial species. The former represents a genomic variable and the latter an environmental variable (a G variable and an E variable in Fig. 2.1). I will first numerically illustrate the conventional method of independent contrasts based on the random-walk Brownian motion model (Felsenstein 1985, 2004, pp. 432–459) and the associated statistical methods for assessing the relationship between two variables. One shortcoming of this method is that the ancestral states of the variables lie somewhere between those of descendent lineages. However, evolution often proceeds with directional changes. For example, various mammalian lineages have in general increased their body size from their humble beginning of tiny insectivores. The ancestral state of the body size, estimated with the assumption of the random-walk Brownian model would be substantially greater than the true one. This shortcoming can be accommodated by the generalized least-square method (Martins and Hansen 1997; Pagel 1997, 1999). The generalized least-square method has an implicit assumption that, if a variable such as body size has exhibited directional change, e.g., having increased in body size in evolutionary lineages, then longer branches should be associated with greater body size. This assumption may not be true because the tree and branch lengths are typically derived from molecular data that do not have direct link to specific phenotypic characters. In addition, the assumption also leads to the restriction that the method for assessing directional change cannot be used with ultrametric trees, i.e., trees with leaves having equal distance to the root such as trees built with a molecular clock. I present an extension of the method of independent contrast, based on the least-square method, to accommodate directional change without this assumption.

Studies of the variation in genomic GC% among bacterial species serve as the easiest entry point into comparative genomics because one does not need any biological knowledge to comprehend the meaning of the variable. Wide variation in genomic GC% is observed in bacterial species. A popular selectionist hypothesis is that bacterial species living in high temperature should have high genomic GC% for two reasons. First, an increased GC usage, with more hydrogen bonds between the two DNA strands, would stabilize the physical structure of the genome (Kushiro et al. 1987; Saenger 1984). Second, high temperature would need more thermostable amino acids (Argos et al. 1979) which are typically coded by GC-rich codons. Such a hypothesis predicts that genomic GC% should increase with optimal grow temperature (OGT) in bacterial species.

The prediction above, however, is not supported empirically. A bacterial species, *Pasteurella multocida*, was cultured under increasing temperatures for ~14,400 generations. GC% was estimated for the ancestral and derived strains by probing both with many AT-rich and GC-rich RAPD primers. If the derived strain has increased genomic GC% during this period of adaptation to increased culture temperature, one would expect to observe more amplification of the GC-rich primers and fewer amplification of AT-rich primers in the derived strain than in the ancestral strain. However, the opposite was observed (Xia et al. 2002). A comparative sequence analysis (Galtier and Lobry 1997) also does not support the prediction.

Surprisingly, it has been found that GC% of rRNA genes is highly correlated with OGT (Dalgaard and Garrett 1993, p. 535; Galtier and Lobry 1997; Hurst and Merchant 2001; Nakashima et al. 2003; Wang and Hickey 2002). In particular, when the loop and stem regions of rRNA are studied separately, it was found that the hyperthermophilic bacterial species not only have higher proportion of GC in the stems but also longer stems (Wang et al. 2006). In contrast, the GC% in the loop region correlates only weakly with OGT. Because stems function to stabilize the RNA secondary structure which is functionally important, these results are consistent with the hypothesized selection for RNA structural stability in high environmental temperatures.

The Necessity of Phylogeny-Based Comparative Method

When studying the relationship between two quantitative variables, such as OGT and stem GC%, a phylogeny-based comparison is crucially important to avoid violation of statistical assumptions. Figure 2.2 illustrates a case in which one may mistakenly conclude a positive relationship between X and Y when the 16 data points are taken as independent. A phylogenetic tree superimposed on the points allows us to see immediately that the data points are not independent. All eight points in the left share one common ancestor, so do the eight points in the right. So the superficial association between X and Y could be due to a single coincidental change in X and Y in one of the two common ancestors. One needs to use the

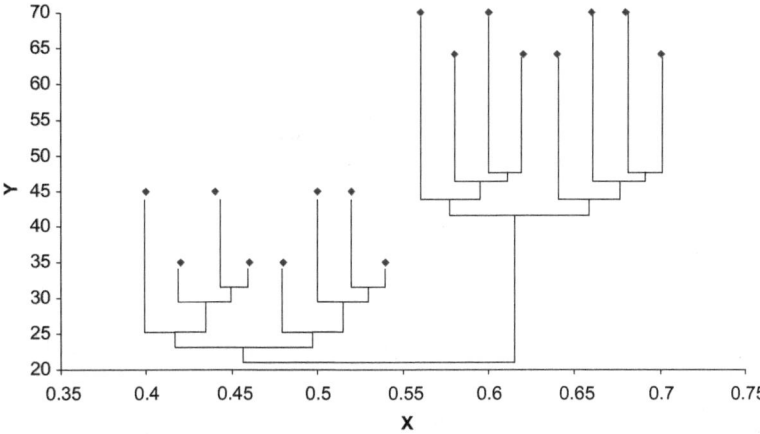

Fig. 2.2 Phylogeny-based comparison is important for evolutionary studies. The data points, when wrongly taken as independent, would result in a significant positive but spurious relationship between Y and X (which represent any two continuous variables, e.g., GC% and OGT)

phylogeny-based method, such as independent contrasts (Felsenstein 1985; 2004, pp. 432–459) or the generalized least-square method (Martins and Hansen 1997; Pagel 1997, 1999) when assessing the relationship between quantitative variables.

While the derivation and mathematical justification of the phylogeny-based comparative method is quite complicated, the most fundamental assumption is the Brownian motion model (Felsenstein 2004, pp. 391–414) which appears reasonable for neutrally evolving continuous characters assumed by the null hypothesis. Here I illustrate the actual computation of independent contrasts with a numerical example to facilitate its application to comparative genomics, prompted by my personal belief that one generally cannot interpret the results properly if one does not know how the results are obtained.

Computing the Independent Contrasts

Suppose a phylogeny of eight bacterial species whose OGT and GC% of rRNA genes have been measured, with the eight species referred to hereafter as s_1 to s_8 from left to right in Fig. 2.3. The computation is recursive, and is exactly the same for any quantitative variable. So we will only illustrate the computation involving OGT. One may repeat the computation involving GC% as an exercise.

The computation is of three steps. First, we recursively compute the ancestral values for internal (ancestral) nodes x_1 to x_6. We treat these ancestors as if they were new taxa and compute the branch lengths leading to these ancestral nodes. We may start with the two sister species s_1 and s_2. The OGT of their ancestor (x_1)

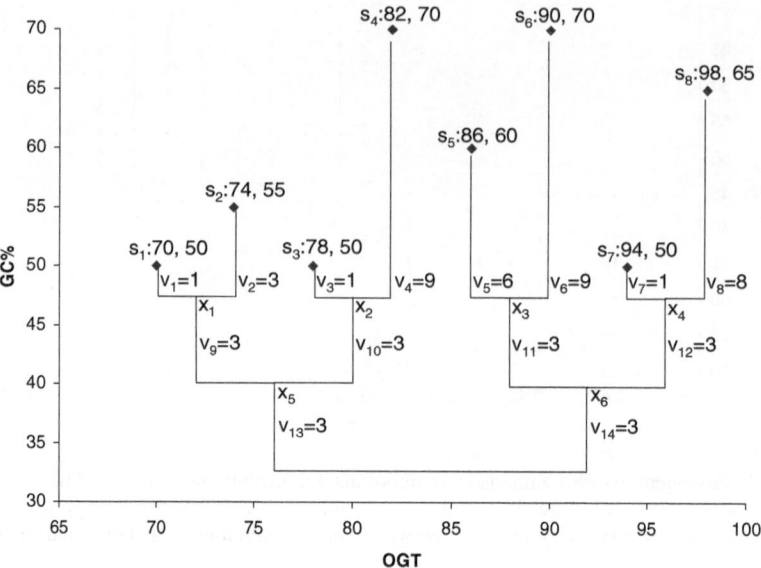

Fig. 2.3 A phylogeny of eight bacterial species (s_1 to s_8) each labeled with optimal growth temperature (OGT) and GC% of the stem region of rRNA genes in the format of "OGT, GC%". The branch lengths (v_1–v_{14}) are next to the branches. Ancestral nodes are designated by x_1 to x_6

is a weighted average of the OGT values for s_1 and s_2 (weighted by the branch lengths):

$$OGT_{x_1} = \frac{v_2}{v_1 + v_2}OGT_{s_1} + \frac{v_1}{v_1 + v_2}OGT_{s_2} = \frac{3 \times 70}{4} + \frac{1 \times 74}{4} = 71 \quad (2.3)$$

One may note that the weighting scheme in Eq. (2.3) is such that the ancestral state is more similar to the state of the descendent node with a shorter branch than the other with a longer branch. This makes intuitive sense as a descendent node diverged much from the ancestor should be less reliable for inferring the ancestral state than a descendent node diverged little from the ancestor.

We now treat x_1 as if it is a new taxon and compute the branch lengths leading to it from its ancestor (x_5) as

$$v_{x_1} = \frac{v_1 v_2}{v_1 + v_2} + v_9 = \frac{1 \times 3}{1 + 3} + 3 = 3.75 \quad (2.4)$$

We do the same for x_2 to x_4, and the associated OGT_{xi} and v_{xi} values are listed in Table 2.1. The computation of the ancestral states for x_5 and x_6 is similar to that in Eq. (2.3), e.g.,

$$OGT_{x_5} = \frac{v_{x_2}OGT_{x_1}}{v_{x_1} + v_{x_2}} + \frac{v_{x_1}OGT_{x_2}}{v_{x_1} + v_{x_2}} = \frac{3.9 \times 71}{7.65} + \frac{3.75 \times 78.4}{7.65} = 74.63$$
$$(2.5)$$

Now we can take the second step to compute the unweighted contrasts (designated by C) as well as the sum of branch lengths linking the two contrasted taxa. With eight species, we have seven ($=n-1$, where n is the number of species) contrasts (first column in Table 2.2). These unweighted contrasts, as well as the sum of branch lengths (SumV) associated with the contrasts, are illustrated for those between s_1 and s_2 and between x_1 and x_2 for OGT in Eq. (2.6). All the computed unweighted contrasts for both OGT and GC%, as well as the associated SumV values, are listed in columns 2–4 in Table 2.2.

$$
\begin{aligned}
C_{s_1 - s_2} . _{OGT} &= OGT_{s_1} - OGT_{s_2} = 70 - 74 = -4 \\
SumV_{C_{s_1 - s_2}} &= v_1 + v_2 = 1 + 3 = 4 \\
C_{x_1 - x_2} . _{OGT} &= OGT_{x_1} - OGT_{x_2} = 71 - 78.4 = -7.4 \\
SumV_{C_{x_1 - x_2}} &= v_{x_1} + v_{x_2} = 3.75 + 3.9 = 7.65
\end{aligned}
\tag{2.6}
$$

We can now take the third step of obtaining independent weighted contrasts (WC) by dividing each unweighted contrasts by the square root of the associated SumV. For example,

$$
\begin{aligned}
WC_{s_1 - s_2} . _{OGT} &= \frac{C_{s_1 - s_2} . _{OGT}}{\sqrt{SumV_{C_{s_1 - s_2}}}} = \frac{-4}{\sqrt{4}} = -2 \\
WC_{x_1 - x_2} . _{OGT} &= \frac{C_{x_1 - x_2} . _{OGT}}{\sqrt{SumV_{C_{x_1 - x_2}}}} = \frac{-7.4}{\sqrt{7.65}} = -2.6755
\end{aligned}
\tag{2.7}
$$

Table 2.1 Computed ancestral states (OGT_{xi} and GC_{xi}) and the branch lengths (v_{xi}) for the six ancestral nodes

x_i	OGT_{xi}	v_{xi}	GC_{xi}
x_1	71.0000	3.7500	51.2500
x_2	78.4000	3.9000	52.0000
x_3	87.6000	6.6000	64.0000
x_4	94.4444	3.8889	51.6667
x_5	74.6275	4.9118	51.6176
x_6	91.9068	5.4470	56.2394

Table 2.2 Unweighted and weighted contrasts for the two quantitative variables OGT and GC%

| Contrast | Unweighted Contrasts | | | Weighted contrasts | |
	OGT	GC%	Sum V	WC_{OGT}	$WC_{GC\%}$
s_1–s_2	−4.0000	−5.0000	4.0000	−2.0000	−2.5000
s_3–s_4	−4.0000	−20.0000	10.0000	−1.2649	−6.3246
s_5–s_6	−4.0000	−10.0000	15.0000	−1.0328	−2.5820
s_7–s_8	−4.0000	−15.0000	9.0000	−1.3333	−5.0000
x_1–x_2	−7.4000	−0.7500	7.6500	−2.6755	−0.2712
x_3–x_4	−6.8444	12.3333	10.4889	−2.1134	3.8082
x_5–x_6	−17.2793	−4.6218	10.3588	−5.3687	−1.4360

These independent contrasts for OGT thus computed, together with those for GC%, are shown in the last two columns in Table 2.2. Now we need to assess the relationship between WC_{OGT} and $WC_{GC\%}$, specifically whether an increase in OGT will result in an increase in GC%, i.e., whether the two are positively correlated. There are two ways to assess the relationship. The first is parametric by performing a linear regression of $WC_{GC\%}$ on WC_{OGT}, forcing the intercept equal to 0. The reason for a zero intercept is that we do not expect a change in GC% if there is no change in OGT. The resulting slope is 0.4647. The regression accounts for 11.17 % of the variation in $WC_{GC\%}$. The square root of 11.17 %, equal to 0.3342, is the correlation coefficient between the two. Of course you may also do a regression of WC_{OGT} on $WC_{GC\%}$, which will result in a slope of 0.2403. These slopes and the correlation coefficients are in the default output in the CONTRAST program in PHYLIP (Felsenstein 2002). The relationship between WC_{OGT} and $WC_{GC\%}$, although positive, is not significant (p = 0.4249).

One may also assess the relationship between WC_{OGT} and $WC_{GC\%}$ by using nonparametric tests. For example, we expect half of the (WC_{OGT}, $WC_{GC\%}$) pairs to have the same sign (i.e., both positive or both negative) and the other half to have different signs. We observe six pairs to have the same sign and one pair to have different signs (Table 2.2). So we have

$$\chi^2 = \frac{(6 - 3.5)^2}{3.5} + \frac{(1 - 3.5)^2}{3.5} = 3.5714 \tag{2.8}$$

with one degree of freedom, the relationship is not significant (p = 0.05878).

Although the method of independent contrasts has been available for many years, many studies, even recent ones, still fall into the same trap, as illustrated in Fig. 2.2, of concluding a significant relationship between X and Y without taking the phylogeny into account. A recent claim of a strong relationship between intron conservation and intron number (Irimia et al. 2007) represents one of such studies.

When the method of independent contrasts was applied to the real data to assess the relationship between bacterial OGT and GC% of rRNA stem sequences and between OGT and rRNA stem lengths, the two relationships are both statistically significant (Wang et al. 2006). Thus, the selectionist hypothesis is supported, but it accounts for only a very small fraction of variation in the genomic GC% among bacterial species, which calls for an alternative hypothesis for the variation in genomic GC%.

One shortcoming of the method of independent contrasts is that the value of the ancestral state is always somewhere between the two values of the descendents. This implies that it cannot detect directional changes over time. For example, if the ancestor is small in body size and all descendents have increased in body size over time, then the Brownian motion model assumed by the independent contrast method is no longer applicable. For example, if we label the root as x_7, then the OGT and GC% values for x_7, OGT_{x_7}, are expected to be

$$OGT_{x_7} = \frac{v_{x5}OGT_{x6}}{v_{x5} + v_{x6}} + \frac{v_{x6}OGT_{x5}}{v_{x5} + v_{x6}} = \frac{4.9118 \times 91.9068}{4.9118 + 5.4470} + \frac{5.4470 \times 74.6275}{4.9118 + 5.4470} = 82.8208$$

$$GC_{x_7} = \frac{v_{x5}GC_{x6}}{v_{x5} + v_{x6}} + \frac{v_{x6}GC_{x5}}{v_{x5} + v_{x6}} = \frac{4.9118 \times 56.2394}{4.9118 + 5.4470} + \frac{5.4470 \times 51.6176}{4.9118 + 5.4470} = 53.8091$$

$$\tag{2.9}$$

However, if we actually know that the ancestral values of OGT and GC% at x_7 are 40 and 45 %, respectively, then these values obviously deviates much from the Brownian expectation. A well known example is the body size of modern mammals which has in general increased substantially from that of the ancestral insectivores since the time of dinosaurs. The Brownian model would lead to the inference of an ancestral body size much larger than that of real insectivore ancestor. It is therefore essential for us to incorporate the known ancestral state to improve the inference.

The generalized least-square method (Martins and Hansen 1997; Pagel 1997, 1999) can be used to accommodate directional changes. However, the method has the limitation that it cannot work with altrametric trees or trees with little variation among the leaf-to-root distances.

Here I present a simple least-square framework to incorporate the ancestral information in the estimation of the values at nodes x_1 to x_6. The residual sum of squares for variable OGT (RSS_{OGT}) is specified below:

$$RSS_{OGT} = \left[40 - \left(\frac{v_{x5}OGT_{x6}}{v_{x5} + v_{x6}} + \frac{v_{x6}OGT_{x5}}{v_{x5} + v_{x6}} \right) \right]^2$$
$$+ \left[OGT_{x6} - \left(\frac{v_{x3}OGT_{x4}}{v_{x3} + v_{x4}} + \frac{v_{x4}OGT_{x3}}{v_{x3} + v_{x4}} \right) \right]^2 + \left[OGT_{x5} - \left(\frac{v_{x1}OGT_{x2}}{v_{x1} + v_{x2}} + \frac{v_{x2}OGT_{x1}}{v_{x1} + v_{x2}} \right) \right]^2$$
$$+ \left[OGT_{x4} - \left(\frac{v_7 OGT_{s8}}{v_7 + v_8} + \frac{v_8 OGT_{s7}}{v_7 + v_8} \right) \right]^2 + \left[OGT_{x3} - \left(\frac{v_5 OGT_{s6}}{v_5 + v_6} + \frac{v_6 OGT_{s5}}{v_5 + v_6} \right) \right]^2$$
$$+ \left[OGT_{x2} - \left(\frac{v_3 OGT_{s4}}{v_3 + v_4} + \frac{v_4 OGT_{s3}}{v_3 + v_4} \right) \right]^2 + \left[OGT_{x1} - \left(\frac{v_1 OGT_{s2}}{v_1 + v_2} + \frac{v_2 OGT_{s1}}{v_1 + v_2} \right) \right]^2$$

$$(2.10)$$

where 40 is the known ancestral value of OGT at the root, and the terms inside the parentheses are the expected OGT values as illustrated before, e.g., Eq. (2.3). To obtain the least-square estimates of OGT values at internal nodes x_1 to x_6, we take the partial derivatives of RSS_{OGT} with respect to OGT_{x1}, OGT_{x2}, ..., OGT_{x6}, set them to zero and solve the six resulting simultaneous equations. The new estimated values of OGT at x_1 to x_6 are 64.3717, 71.0352, 85.7588, 86.9876, 53.8856, and 77.4462, respectively. The new values suggest that the OGT values have increased in all descendent lineages from the ancestral value of 40, i.e., the OGT values at ancestral nodes are consistently smaller than the descendent lineages.

The least-square frame is not limited to one known ancestral value. For example, if OGT_{x5} is known, it can be substituted into Eq. (2.10) so that we will only need to estimate five unknown ancestral OGT values. The same computation can be done for GC% or any other variable with one or more known ancestral values. The independent contrasts can be computed the same way as before, except that the new ancestral values are then used.

The mutation hypothesis of genomic GC% variation (Muto and Osawa 1987; Sueoka 1964; Xia and Yuen 2005; Xia et al. 2002) invokes biased mutation in different bacterial species to explain genomic variation in GC%, i.e., GC-rich genomes are the result of GC-biased mutation. One prediction from the mutation hypothesis is that the third codon position should increase more rapidly with the genomic GC% than the first codon position which in turn should have its GC% increase more rapidly with the genomic GC% than the second codon position. The reason for this prediction is that the third codon positions are little constrained functionally because most substitutions at the third codon positions are synonymous. Some nucleotide substitutions at the first codon positions are synonymous, but most are nonsynonymous. All nucleotide substitutions at the second codon positions are nonsynonymous and typically involve rather different amino acids (Xia 1998b; Xia and Li 1998). The empirical results (Fig. 2.4) strongly support the prediction above (Muto and Osawa 1987).

However, the pattern in Fig. 2.4, while consistent with the mutation hypothesis, has resulted in two misconceptions. First, the pattern shown by the third codon position is often interpreted to reflect mutation bias. This interpretation is incorrect because the third codon position is subject to selection by differential availability of tRNA species (Carullo and Xia 2008; Xia 1998a, 2005, 2008; Xia et al. 2007). We may contrast a GC-rich *Streptomyces coelicolor* and a GC-poor *Mycoplasma capricolum* as an illustrative example. *M. capricolum* has no tRNA with a C or G at the wobble site for four-fold codon families (Ala, Gly, Pro, Thr and Val), i.e., the translation machinery would be inefficient in translating C-ending or G-ending

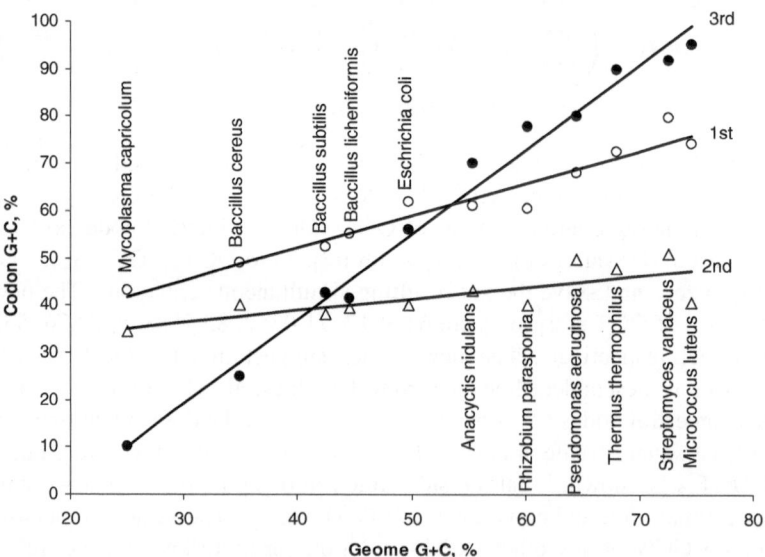

Fig. 2.4 Correlation of GC% between genomic DNA and first, second and third codon positions (Muto and Osawa 1987). While the actual position of the points may be substantially revised with new genomic data (e.g., the GC% for the first, second and third codon positions for *Mycoplasma capricolum* is 35.8, 27.4, and 8.8 % based on all annotated CDSs in the genomic sequence), the general trend remains the same

codons. This implies selection in favour of A-ending or U-ending codons and will consequently reduce GC% at the third codon position. This most likely has contributed to the low GC% at the third codon position in *M. capricolum*. In contrast, most of the tRNA genes translating the five four-fold codon families in the GC-rich *S. coelicolor* have G or C at the wobble site, and should favour the use of C-ending or G-ending codons. This most likely has contributed to the high GC% at the third codon position in *S. coelicolor*. The same pattern is observed for two-fold codon families. The most conspicuous one is the Gln codon family (CAA and CAG). There is only one tRNA^{Gln} gene in *M. capricolum* with a UUG anticodon favouring the CAA codon. In contrast, there are two tRNA^{Gln} in *S. coelicolor*, both with a CUG anticodon favouring the CAG codon. Thus, the high slope for the third codon position in Fig. 2.4 is at least partially attributable to the tRNA-mediated selection. Relative contribution of mutation and tRNA-mediated selection to codon usage has been evaluated in several recent studies (Carullo and Xia 2008; Xia 2005, 2008; Xia et al. 2007).

Second, the observation that GC% of the third codon position increases with genomic GC% is sometimes taken to imply that the frequency of G-ending and C-ending codons will increase with genomic GC% or GC-biased mutation (Kliman and Bernal 2005). This is not generally true. Take the arginine codons for example. Given the transition probability matrix for the six synonymous codons shown in Table 2.3, the equilibrium frequencies (π) for the six codons are

$$\pi_{AGA} = \frac{1}{2k^2 + 3k + 1}$$

$$\pi_{AGG} = \pi_{CGA} = \pi_{CGT} = \frac{k}{2k^2 + 3k + 1} \quad (2.11)$$

$$\pi_{CGC} = \pi_{CGG} = \frac{k^2}{2k^2 + 3k + 1}$$

The three solutions correspond to the number of GC in the codon, with AGA having one, AGG, CGA and CGT having two, and CGC and CGG having three

Table 2.3 Transition probability matrix for the six synonymous arginine codons, with α for transitions (C$\leftrightarrow$T and A$\leftrightarrow$G), β for transversions, and k modeling AT-biased mutation ($0 \leq k \leq 1$) or GC-biased mutation ($k > 1$)

	CGT	CGC	CGA	CGG	AGA	AGG
CGT		kα	β	kβ	0	0
CGC	α		β	β	0	0
CGA	β	kβ		kα	β	0
CGG	β	β	α		0	β
AGA	0	0	kβ	0		kα
AGG	0	0	0	kβ	α	

We ignore nonsynonymous substitutions because nonsynonymous substitution rate is often negligibly low compared to synonymous rate. The diagonal is constrained by the row sum equal to 1

G or C. One may note that the G-ending codon AGG has the same equilibrium frequency as that of the A-ending CGA and the T-ending CGT. Thus, we should not expect A-ending or T-ending codons to always decrease, or G-ending and C-ending codons always increase, with increasing genomic GC% or GC-biased mutation. In fact, according to the solutions in Eq. (2.11), π_{AGG}, π_{CGA}, and π_{CGT} will first increase with k until k reaches $\sqrt{2}/2$, and will then decrease with k when $k > \sqrt{2}/2$ (Palidwor et al. 2010).

One may ask why the phylogeny-based comparison was not used for characterizing the relationship between codon GC% and genomic GC% in the 11 species in Fig. 2.4. The reason is that the two variables change very fast relative to the divergence time among the studied species, i.e., phylogenetic relatedness among the 11 species is a poor predictor of the codon GC% or genomic GC%. That genomic GC% has little phylogenetic inertia is generally true in prokaryotic species (Xia et al. 2006). In such cases, one may assume approximate data independence and perform a phylogeny-free analysis. Another study that leads to insight into the relationship between UV exposure and GC% in bacterial genomes (Singer and Ames 1970), which may be the first comparative genomic study, is also not phylogeny-based.

The Comparative Methods for Discrete Characters

A genome typically encodes many genes. The presence or absence of certain genes, certain phenotypic traits and environmental conditions jointly represent a major source of data for comparative genomic analysis. These binary data are best analyzed by comparative methods for discrete data.

A total of 11728 bacterial genomes and 249 archaea genomes have been made available for research through Entrez as of May 29, 2012. In addition to genomic GC that can be computed as soon as the sequences are available, each sequencing project also delivers a list of genes in the sequenced genome, identified by one of two categories of methods, i.e., by checking against the "gene dictionary" through homology search, e.g., BLAST (Altschul et al. 1990, 1997) or by computational gene prediction, e.g., GENSCAN (Burge and Karlin 1997, 1998). The availability of such annotated genomes, as well as the availability of powerful phylogenetic software packages (Aris-Brosou and Xia 2008) such as MEGA (Kumar et al. 2008), PAUP* (Swofford 2000), PHYLIP (Felsenstein 2002), BEAST (Drummond and Rambaut 2007) and DAMBE (Xia 2001), greatly facilitates the compilation of data for comparative genomics illustrated in Fig. 2.1.

One concrete example is shown in Fig. 2.5. We can study the column variables individually or associations between column variables. For each column of gene presence/absence data, the absence can be attributed to gene loss, but the presence of a gene in a gene may either result from inheritance from the ancestor or from lateral gene transfer which occurs frequently in bacterial species. Phylogeny-based

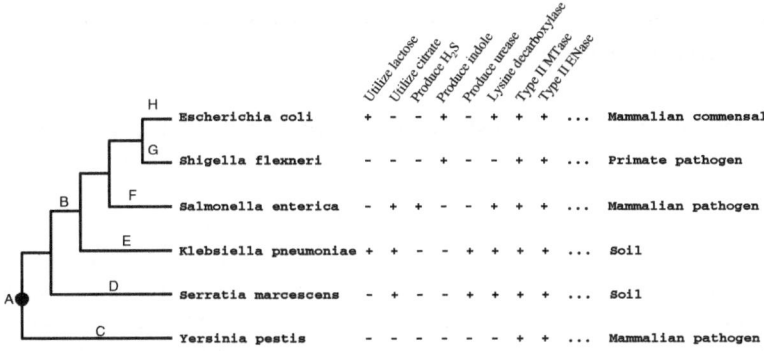

Fig. 2.5 Phylogeny-based comparative bacterial genomics, with ± indicating the presence/absence of gene-mediated functions. Modern bacterial comparative genomics typically would have thousands of columns each representing the presence/absence of one gene function as well as many environmental variables of which only a habitat variable is shown here. Modified from Ochman et al. (2000)

inferences, such as phylogenetic incongruence test illustrated in the next section, help us identify genes that tend to be laterally transferred. It is the discovery of the rampant occurrence of lateral gene transfer that lead to the realization that the cenancestor is neither a single cell nor a single genome, but is instead an entangled bank of heterogeneous genomes with relatively free flow of genetic information. Out of this entangled bank of frolicking genomes arose probably many evolutionary lineages with gradually reduced rate of horizontal gene transfer confined mainly within individual lineages (Xia and Yang 2012). Only three (Archaea, Eubacteria and Eukarya) of these early lineages have representatives survived to this day.

Identifying laterally transferred genes is not only important in its own merit, but also crucial in molecular phylogenetics for building species trees which should rely only on ancestrally inherited characters. The late Ernst Mayr, once in an argument against using parasites as markers to infer phylogeny in a conference, stated that two birds can exchange parasites but never exchange their heads or wings or legs (Paterson et al. 1995). The point is that we should use characters such as heads, wings, and tails that are ancestrally inherited instead of parasites that could be laterally transferred to build phylogenetic relationships. Phylogeny-based comparative methods require accurate phylogeny. It is mainly for this reason that the phylogenetic incongruence test was detailed in the next section.

While studying individual genes has its merits, comparative genomes is mainly about association between genes and between genes and phenotypic and environmental variables. The phylogeny-based comparative method (Barker and Pagel 2005; Pagel 1994) for characterizing such associations for discrete variables is also numerically illustrated in this section.

Studying Variables Individually: Detecting Genes that Tend to be Laterally Transferred

While there are many ways to study the variables individually, here we illustrate only one type of study, i.e., detecting lateral gene transfer (LGT). We may focus on the first column in Fig. 2.5. First, *Escherichia coli* and *Klebsiella pneumoniae* have genes coding proteins for lactose metabolism, but others do not. This leads to at least three possible evolutionary scenarios. First, lactose-metabolizing function may be absent in the ancestor A (Fig. 2.5), but (1) gained along lineage B and lost in lineage F and G or (2) gained independently along lineage E and lineage H (e.g., by LGT). The third possible scenario is that the function is present in the ancestor A, but lost in all species except for lineages E and H.

If lactose-metabolizing genes are frequently involved in LGT, then we should expect the gene tree built from the lactose operon genes to be different from the species tree which is typically approximated by a tree built from many housekeeping genes. Is the lactose operon gene tree significantly different from the species tree?

Suppose we have the sequence data (Fig. 2.6) from housekeeping genes, a species tree (T_1) and a lactose operon gene tree (T_2). We wish to test whether T_1 is significantly better than T_2 given the housekeeping gene sequences, with the null hypothesis being that T_2 is just as good as T_1. Both the maximum parsimony (MP) and the maximum likelihood (ML) methods have been used for such significance tests.

For the ML method, we compute the log-likelihood (lnL) for each of the nine sites (Fig. 2.6) given T_1 and T_2, respectively (lnL$_1$ and lnL$_2$ for T_1 and T_2, respectively, Table 2.4). A simple numerical illustration of computing site-specific lnL can be found in Xia (2007a, pp. 279–280). A paired-sample *t* test can then be applied to test whether mean lnL$_1$ is significantly different from mean lnL$_2$. For our data in Table 2.4, t = 4.107, DF = 8, p = 0.0034, two-tailed test). So we reject the null hypothesis and conclude that the lactose operon gene tree (T_2) is significantly worse than the species tree (T_1). A natural explanation for the phylogenetic incongruence is LGT.

For the MP method, we compute the minimum number of changes (NC) for each site given T_1 and T_2 (Fig. 2.6), respectively (NC$_1$ and NC$_2$ for T_1 and T_2,

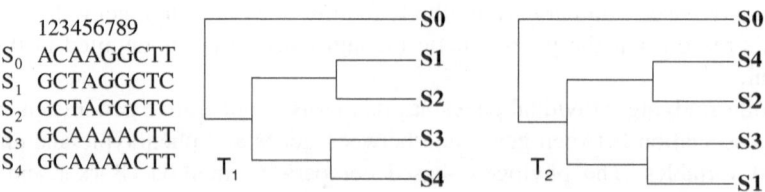

Fig. 2.6 DNA sequence data for significance tests of two alternative topologies

Table 2.4 Phylogenetic incongruence tests with maximum likelihood (ML) and maximum parsimony (MP) methods. $\ln L_1$ and $\ln L_2$ are site-specific log-likelihood values based on the F84 model and T_1 and T_2 (Fig. 2.6), respectively, and NC1 and NC2 are the minimum number of changes required for each site given T_1 and T_2, respectively

Site	ML		MP	
	$\ln L_1$	$\ln L_2$	NC_1	NC_2
1	−4.0975	−4.0990	1	1
2	−2.0634	−2.7767	0	0
3	−5.1147	−7.7335	1	2
4	−1.9481	−2.6238	0	0
5	−3.2142	−5.0875	1	2
6	−3.2142	−5.0875	1	2
7	−2.0634	−2.7767	0	0
8	−2.3938	−3.2626	0	0
9	−3.1090	−3.8572	1	2

respectively, Table 2.4). A simple numerical illustration of computing site-specific NC can be found in Xia (2007a, pp. 272–275). We can then perform a paired-sample t test as before to test whether mean NC_1 is significantly smaller than NC_2, in one of three ways. The first is to use the entire nine pairs of data, which yields $t = −2.5298$, DF $= 8$, $p = 0.0353$, and a decision to reject the null hypothesis that T_1 and T_2 are equally good at the 0.05 significance level, i.e., T_1 is significantly better than T_2. Second, we may use only the five polymorphic sites in the paired-sample t test, which would yield $t = −4$, DF $= 4$, and $p = 0.0161$. This leads to the same conclusion. The third is to use only the four informative sites which is however inapplicable in our case because we would have four NC_1 values all equal to 1 and four NC_2 values all equal to 2, i.e., the variation in the difference is zero.

When the phylogenetic incongruence test is applied to real lactose operon data, it was found that the lactose operon gene tree is somewhat compatible to the species tree, and the case for LGT is therefore not strong (Stoebel 2005). This suggests the possibility that the lactose operon was present in the ancestor, but has been lost in a number of descendent lineages. In contrast, the urease gene cluster, which is important for long-term pH homeostasis in the bacterial gastric pathogen, *Helicobacter pylori* (Sachs et al. 2003; Xia and Palidwor 2005), generate genes trees significantly different from the species tree (unpublished result). This suggests that the urease gene cluster is involved in LGT and has implications in emerging pathogens. For example, many bacterial species pass through our digestive system daily, and it is conceivable that some of them may gain the urease gene cluster and become acid-resistant, with the consequence of one additional pathogen for our stomach.

One may note that significant incongruence between the gene tree and species tree does not imply LGT because events such as gene duplication and lineage-specific gene loss can also lead to phylogenetic incongruence (Page 2003). This is illustrated in Fig. 2.7 with five species labelled Sp1 to Sp5. A gene duplication

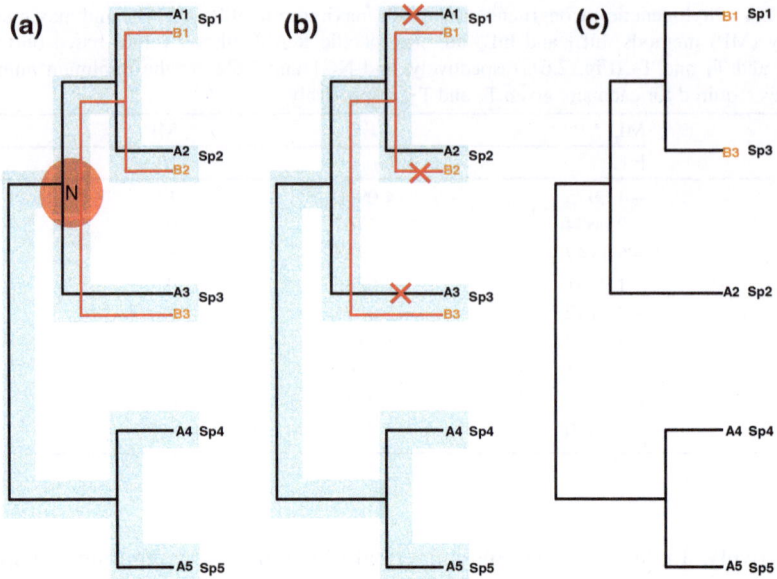

Fig. 2.7 Phylogenetic incongruence can result from gene duplication and lineage-specific gene loss. **a** A gene duplication event occurred at Node N. **b** Genes A1, B2 and B3 were lost in evolution. **c** Phylogenetic tree resulting from the remaining 5 gene sequences is different from the true tree (*shaded*)

event occurred at node N in Fig. 2.7a, leading to paralogous genes A and B in all subsequent lineages. Differential gene losses occurred subsequently (Fig. 2.7b), leading to the loss of A1, B2 and A3, which would mislead us to think that gene duplication has never occurred and the gene has always been in a single-copy state. Using these five gene sequences, B1, B3, A2, A4 and A5, we would arrive at the wrong tree that is different from the true tree in Fig. 2.7a. However, genes that have undergone such duplications and lineage-specific gene losses are also poor phylogenetic markers. The illustration does not invalidate the use of the phylogenetic incongruence test to identify genes that are poor phylogenetic markers.

Studying Association Between Variables

Many genes work together and complement each other to accomplish a biological function. For example, Type II ENase (restriction endonuclease) is always accompanied by the same type of MTase (methyltransferase) recognizing the same site (Fig. 2.5). Patterns like this allow us to quickly identify enzymes that are partners working in concert. ENase cuts the DNA at specific sites and defends the bacterial host against invading DNA phages. MTase modifies (methylates) the same site in the bacterial genome to prevent ENase from cutting the bacterial genome. Obviously, ENase activity without MTase is suicidal, so MTase must accompany ENase, although ENase may get lost without immediate detrimental effect. The functional complementation also explains why the activity of many ENases depends on S-adenosylmethionine (AdoMet) availability. AdoMet always serves as the methyl donor for MTase. Without AdoMet, the restriction sites in the host genome will not be modified even in the presence of MTase because of the lack of the methyl donor, and ENase activity will then kill the host. So it is selectively advantageous for ENase activity to depend on the availability of AdoMet. Although rare, MTase can be present without the associated ENase. For example, *E. coli* possesses two unaccompanied MTases, Dam and Dcm. Some bacteriophages carry one or more MTases to modify their own genome so as to nullify the hostile action of the host ENases.

Aside from association between genes, we are often interested in the association between gene function and environmental variables. For example, the production of functional urease is often associated with an acidic environment in bacterial species, such as *Helicobacter pylori*, *Klebsiella pneumoniae* and *Serratia marcescens*. *H. pylori* inhabits the acidic environment in mammalian stomach, and the two other species can generate acids by fermentation leading to acidification of their environment. The presence of urease, which catalyzes urea to produce ammonia, can help maintain cytoplasmic pH homeostasis and allow them to tolerate environmental pH of 5 or even lower. Thus, comparative genomics can help us understand gene functions in particular environmental conditions.

Urease gene cluster serves as one of the two key acid-resistant mechanisms in the bacterial pathogen *Helicobacter pylori* in mammalian stomach, with the other mechanism being a positively charged cell membrane that alleviates the influx of protons into cytoplasm. The latter mechanism is established by comparative genomics between *H. pylori* and its close relatives as an adaptation to the acidic environment in the mammalian stomach (Xia and Palidwor 2005).

Not only can association between genes, or between genes and their function can lead to biological insights, but the lack of certain expected association can also shed light on gene functions. For example, a set of *ERG* genes involved in *de novo* cholesterol biosynthesis are strongly conserved among various animal lineages. However, some of these genes are also strongly conserved in *Drosophila melanogaster* and *Caenorhabditis elegans* that are unable to synthesize cholesterol, i.e., a de-coupling of the genes and their expected function. Comparative genomics studies suggest that the *ERG* homologs in *D. melanogaster* and *C. elegans* have evolved to acquire new functions (Vinci et al. 2008).

The identification of association either between two genes (e.g., between a type II ENase and a type II MTase) or between a gene and an environmental variable (e.g., between urease production and the habitat) represents the same statistical problem. However, a statistician without biological background may misconstrue the problem and might use a 2 × 2 contingency table (i.e., $N_{+/+}$, $N_{+/-}$, $N_{-/+}$, $N_{-/-}$) and Fisher's exact test to identify the association between two columns without taking the phylogeny into consideration. However, such an approach can

lead to both false negatives and false positives. Figure 2.8 illustrates the association study of two pairs of genes. Ignoring the phylogeny will lead to a significant
association between genes *ORC3* and *CIN3*. However, the data points are not independent as the superficial association could be caused by only two consecutive
gene-gain events (Fig. 2.8) and all the seven "11" could then be the consequence
of shared ancestral characters.

A phylogeny-based comparative analysis (Barker and Pagel 2005; Pagel 1994)
characterizes the state transition by a Markov chain, and uses a likelihood ratio
test to detect the presence of association between genes or between a gene function and an environmental condition. Two genes, each with two states (presence/
absence), have four possible joint states and eight rate parameters (α_1, α_2, β_1, β_2,
δ_1, δ_2, γ_1 and γ_2) to be estimated from the data (Fig. 8). When the gain or loss of
one gene is independent of the other gene, then $\alpha_1 = \alpha_2$, $\beta_1 = \beta_2$, $\delta_1 = \delta_2$, and
$\gamma_1 = \gamma_2$, with only four rate parameters to be estimated. Thus, we compute the
log-likelihood for the eight-parameter and four-parameter model given the tree and

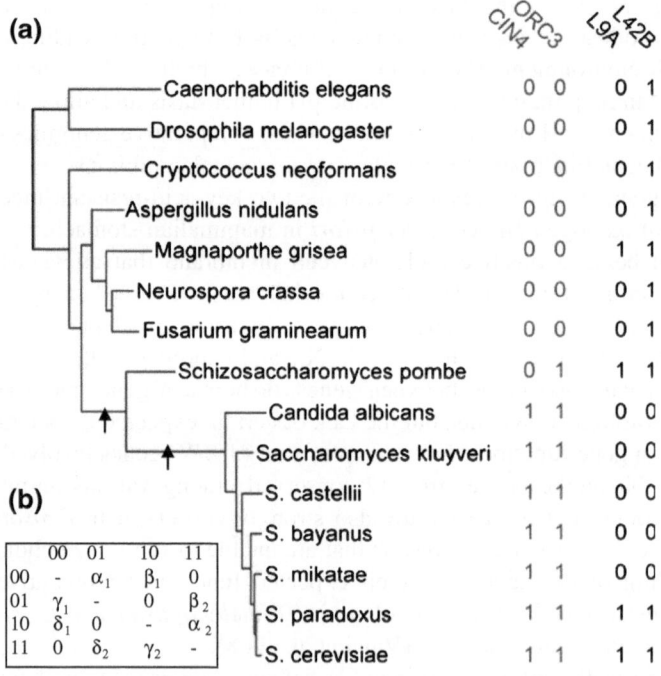

Fig. 2.8 Comparative methods for discrete binary characters. The presence and absence (designated by 1 and 0, respectively) of four genes are recorded for each species (**a**) The two *black
arrows* indicate a gene-gain event. The instantaneous rate matrix (**b**), with notations following
Felsenstein (2004), shows the relationship among the four character designation, i.e., 00 for both
genes absent, 01 for the absence of gene 1 but presence of gene 2, 10 for the presence of gene 1
but absence of gene 2, and 11 for both genes present. The diagonals are constrained by each row
sum equal to 0. Modified from (Barker and Pagel 2005)

the data, designated $\ln L_8$ and $\ln L_4$, respectively, and perform a likelihood ratio test with test statistic being $2(\ln L_8 - \ln L_4)$ and four degrees of freedom.

I illustrate the computation of $\ln L_8$ by using a simpler tree with only four operational taxonomic units or OTUs (Fig. 2.9). The joint states, represented by binary numbers 00, 01, 10 and 11, correspond to decimal numbers 0, 1, 2 and 3 which will be used to denote the four states in some equations below. The likelihood for the eight-parameter model is

$$L_8 = \sum_{z=0}^{3} \sum_{y=0}^{3} \sum_{x=0}^{3} \pi_z P_{zx}(b_6) P_{x0}(b_1) P_{x3}(b_2) P_{zy}(b_5) P_{y0}(b_3) P_{y3}(b_4)$$

(2.12)

Equation (2.12) may seem to suggest that we need to sum 3^4 terms. However, the amount of computation involved is greatly reduced by the pruning algorithm (Felsenstein 1981). To implement this algorithm, we define a vector L with elements $L(0)$, $L(1)$, $L(2)$, and $L(3)$ for every node including the leaves. L for leaf i is defined as

$$L_i(s) = \begin{cases} 1, & \text{if } s = S_i \\ 0, & \text{otherwise} \end{cases}$$

(2.13)

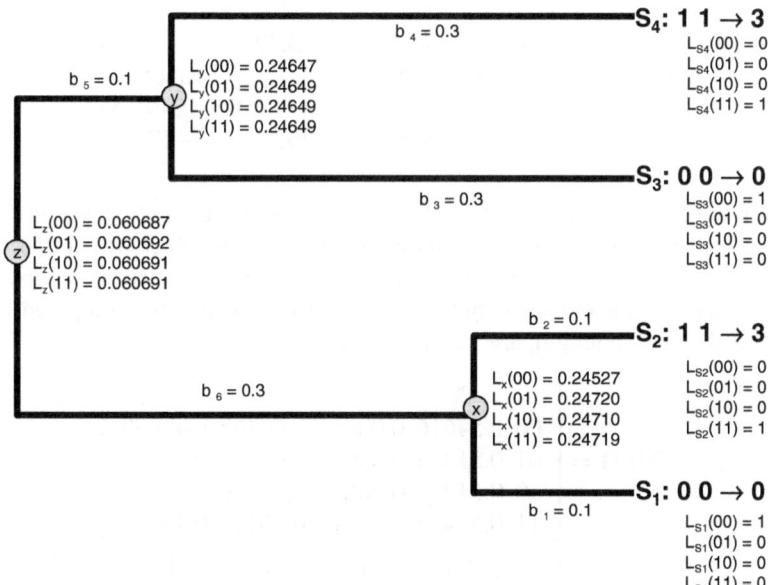

Fig. 2.9 Four-OTU tree with branch lengths (b1 to b6) for illustrating likelihood computation. The L vectors are computed recursively according to Eqs. (2.13)–(2.14)

L for an internal node with two offspring (o_1 and o_2) is recursively defined as

$$L_i(s) = \left[\sum_{k=0}^{3} P_{sk}(b_{i,o_1}) L_{o_1}(k) \right] \left[\sum_{k=0}^{3} P_{sk}(b_{i,o_2}) L_{o_2}(k) \right] \qquad (2.14)$$

where $b_{i,\,o1}$ means the branch length between internal node i and its offspring o_1, and P_{sk} is the transition probability from state s to state k computed from the rate matrix (Fig. 2.8b). For example, $b_{x,\,S1}$ (branch length between internal node x and its offspring S_1) is b_1 in Fig. 2.9. The computation involves finding the eight rate parameters that maximize L_8. As there is no analytical solution, the maximizing algorithm will simply try various rate parameter values and evaluate L_8 repeatedly until we converge on a set of parameter values that result in maximum L_8. Many such algorithms are well explained and readily available in source code (Press et al. 1992).

While the equations might be confusing to some, the actual computation is quite simple. With only four OTUs, $S_1 = S_3 = $ '00' and $S_2 = S_4 = $ '11' (Fig. 2.9), the likelihood surface is quite flat and many different combination of the rate parameters can lead to the same maximum L_8. In fact, the only constraint on the rate parameters is high rates from states 01 and 10 to states 00 and 11 (i.e., large $\delta_1 + \gamma_1 + \alpha_2 + \beta_2$) and low rates from states 00 and 11 to states 01 and 10 (i.e., (i.e., small $\delta_2 + \gamma_2 + \alpha_1 + \beta_1$). This should be obvious when we look at the four OTUs in the tree (Fig. 2.9), with only 00 and 11 being observed at the leaves. This implies that 01 and 10 should be transient states, quickly changing to 00 or 11, whereas 00 and 11 are relatively conservative stable states. One of the rate matrices that approaches the maximum L_8 is

$$Q = \begin{bmatrix} & 00 & 01 & 10 & 11 \\ 00 & -16.47 & 13.15 & 3.32 & 0 \\ 01 & 1.10 & -135653.97 & 0 & 135652.87 \\ 10 & 1816.49 & 0 & -20308.04 & 18491.54 \\ 11 & 0 & 18.30 & 207.21 & -225.52 \end{bmatrix} \qquad (2.15)$$

The rate of transition from states 01 and 10 to states 00 and 11 is 644.5 times greater (The true rate should be infinitely greater) than the other way round, which implies that we will almost never observe 01 and 10 states. The transition probability matrices with branch lengths of 0.1 and 0.3, which are computed as e^{Qt}, where t is the branch length, are, respectively,

$$P(0.1) = \begin{bmatrix} & 00 & 01 & 10 & 11 \\ 00 & 0.54616 & 0.00011 & 0.00467 & 0.44908 \\ 01 & 0.51459 & 0.00011 & 0.00499 & 0.48038 \\ 10 & 0.51738 & 0.00011 & 0.00496 & 0.47759 \\ 11 & 0.51458 & 0.00011 & 0.00499 & 0.48034 \end{bmatrix}$$

$$P(0.3) = \begin{bmatrix} & 00 & 01 & 10 & 11 \\ 00 & 0.53145 & 0.00011 & 0.00482 & 0.46377 \\ 01 & 0.53144 & 0.00011 & 0.00482 & 0.46382 \\ 10 & 0.53144 & 0.00011 & 0.00482 & 0.46382 \\ 11 & 0.53144 & 0.00011 & 0.00482 & 0.46382 \end{bmatrix}$$

$$(2.16)$$

We can now compute L_8 by using the pruning algorithm. First, L_{S1} to L_{S4} are straightforward from Eq. (2.13) and shown in Fig. 2.9. L_x and L_y are computed according to Eq. (2.14), e.g.,

$$
\begin{aligned}
L_x(00) &= P_{0000}(0.1)P_{0011}(0.1) = 0.54616 \times 0.44908 = 0.24527 \\
L_x(01) &= 0.51459 \times 0.48038 = 0.24720 \\
L_x(10) &= 0.51738 \times 0.47759 = 0.24710 \\
L_x(11) &= 0.51458 \times 0.48037 = 0.24719
\end{aligned}
\tag{2.17}
$$

Similarly, $L_y(00)$, $L_y(01)$, $L_y(10)$, and $L_y(11)$ are computed the same way and have values 0.24647, 0.24649, 0.24649, and 0.24649, respectively. Similarly, L_z is also computed by applying Eq. (2.14), e.g.,

$$
\begin{aligned}
L_z(00) = AB &= 0.246207 \times 0.246487 = 0.060687, \; where \\
A &= [P_{0000}(b_6)L_x(00) + P_{0001}(b_6)L_x(01) + P_{0010}(b_6)L_x(10) + P_{0011}(b_6)L_x(11)] \\
&= 0.246207 \\
B &= [P_{0000}(b_5)L_y(00) + P_{0001}(b_5)L_y(01) + P_{0010}(b_5)L_y(10) + P_{0011}(b_5)L_y(11)] \\
&= 0.246487
\end{aligned}
\tag{2.18}
$$

$L_z(01)$, $L_z(10)$, and $L_z(11)$ are 0.060692, 0.060691, and 0.060691, respectively. The final L_8 is

$$
L_8 = \sum_{k=0}^{3} \pi_k L_z(k) = 0.060687 \times 0.5 + 0.060691 \times 0.5 = 0.060689
\tag{2.19}
$$

$$
\ln(L_8) = -2.802
$$

where we used the empirical frequencies for π_k, although π_k could also be estimated as a parameter of the model. Note that states 01 and 10 are not observed, and π_{01} and π_{10} are assumed to be 0 in Eq. (2.19).

The computation of $\ln(L_4)$ is simpler because only four rate parameters need to be estimated, and is equal to -5.545. If quite a large number of OTUs are involved, then twice the difference between the two log-likelihood, designated $2\Delta\ln L$, follows approximately the χ^2 distribution with 4 degrees of freedom. If we could assume large-sample approximation in our case, then $2\Delta\ln L = 5.486$, which leads to p = 0.241, i.e., the eight-parameter model is not significantly better than the four-parameter model. Such a result is not surprising given the small number of OTUs.

With this phylogeny-based likelihood approach, Barker et al. (2007) found that the superficial association between genes *CIN4* and *ORC3* is not significant, although Fisher's exact test ignoring the phylogeny would produce a significant association between the two genes. Similarly, genes *L9A* and *L42B* were found to be significantly associated based on the phylogeny-based likelihood approach, although Fisher's exact test ignoring the phylogeny would suggest a lack of the association. In this particular case, *L9A* and *L42B* are known to be functionally associated and *CIN4* and *ORC3* are known not be functionally associated. Ignoring the phylogeny would have produced both a false positive and a false negative. Phylogeny-based comparative methods for continuous and discrete methods have been implemented in the freely available software DAMBE(Xia 2001; Xia and Xie 2001) at http://dambe. bio.uottawa.ca.

Sometimes one may find the presence of orthologous genes in different species but the function associated with the gene is missing in some species, i.e., the same genotype (presence of a gene or a group of genes) correspond to different phenotypes. Such is the case of *ERG* genes involved in sterol metabolism. Many species, including *Drosophila melanogaster* and *Caenorhabditis elegans*, share orthologous genes involved in *de novo* sterol synthesis (Vinci et al. 2008), but *D. melanogaster* and *C. elegans* have lost their ability to synthesize sterols *de novo*, although their *ERG* orthologs are still under strong purifying selection revealed by a much lower nonsynonymous substitution rate than the synonymous substitution rate. Further microarray studies demonstrated a strong association between the orthologs of *ERG24 and ERG25* in *D. melanogaster* and genes involved in ecdysteroid synthesis and in intracellular protein trafficking and folding (Vinci et al. 2008). This suggests that the *ERG* genes in *D. melanogaster* have evolved new association with other genes and gained new functions.

Mapping genes and gene functions to a phylogeny has revealed the loss of an essential single-copy *Maelstrom* gene in fish, and a plausible explanation is that the essential function has been fulfilled by a non-homologous gene (Zhang et al. 2008). Thus, the same phenotype can have different genotypes. Such findings that a specific molecular function can be performed by evolutionarily unrelated genes suggest a fundamental flaw in research effort to identify the minimal genome by identifying shared orthologous genes (Mushegian and Koonin 1996). The rationale for such an approach is this. Suppose a minimal organism needs to perform three essential functions designated {x, y, z}, and three different genes, designated {*A, B, C*}, encode products that perform these three functions. If we have a genome (G1) with five genes {*A, B, C, D, E*} and another genome (G2) with four genes {*A, B, C, F*}, with genes of the same letter being orthologous, then shared orthologous genes between G1 and G2 are {*A, B, C*} which would be a good approximation of the minimal genome. However, it is possible that G1 = {*A, D, E*} for essential functions {x, y, z} and G2 = {*A, C, F*} for the same set of functions {x, y, z}. Both are already minimal genomes, but the intersection of G1 and G2 is only {*A*} which is a severe underestimation of a minimal genome. Creating a cell with such a "minimal" genome is doomed to fail.

The comparative methods still need further development. For example, one difficulty with the comparative methods for the continuous and discrete characters is what branch lengths to use because different trees, or even the same topology with different branch lengths, can lead to different conclusions. One may need to explore all plausible trees to check the robustness of the conclusion.

Multiple Comparisons and the Method of False Discovery Rate

The material in this section can be found in several books and papers (e.g., Ma and Xia 2011). It is included purely for reader's convenience.

Modern comparative genomic studies may often involve the functional association of thousands of genes or more. With N genes, there are $N(N-1)/2$ possible pairwise associations and $N(N-1)/2$ tests of associations. There are $N(N-1)(N-2)/6$ possible triplet associations. So it is necessary to consider the topic of how to control for error rates in multiple comparisons.

There are two approaches for adjusting type I error rate involving multiple comparisons, one controlling for familywise error rate (FWER), and the other controlling for the false discovery rate (FDR) (Nichols and Hayasaka 2003). While FWER methods are available in many statistical packages and covered in many books, there are few computational tutorials for the FDR in comparative genomics, an imbalance which I will try to compensate below.

The difference between the FDR and FWER is illustrated in Table 2.5, where N_{12} denotes the number of null hypotheses that are true but rejected (false positives). FWER is the probability that N_{12} is greater or equal to 1, whereas FDR is the expected proportion of $N_{12}/N_{.2}$ and defined to be 0 when $N_{.2} = 0$. Thus, FDR is a less conservative protocol for comparison, with greater power than FWER, but at a cost of increasing the likelihood of obtaining type I errors.

The FDR protocol works with a set of p values. For example, with 10 genes, there are 45 pairwise tests of gene associations, yielding 45 p values. The FDR protocol is to specify a reasonable FDR (typically designated by q) and find a critical p (designated $p_{critical}$) so that a p value that is smaller than $p_{critical}$ is considered as significant, otherwise it is not. The q value is typically 0.05 or 0.01. Two general FDR procedures, Benjamini-Hochberg (BH) and Benjamini-Yekutieli (BY), are illustrated below.

Suppose we have a set of 15 sorted p values from testing 15 different hypotheses (Table 2.6). The Bonferroni method uses α/m (where m is the number of p values) as a critical p value ($p_{critical . Benferroni}$) for controlling for FWER. We have $m = 15$. If we take $\alpha = 0.05$, then $p_{critical . Benferroni} = 0.05/15 = 0.00333$ which would reject the first three hypotheses with the three smallest p values.

Table 2.5 Cross-classificationof N tests of hypothesis

H_0	Reject	
	No	Yes
TRUE	N_{11}	N_{12}
FALSE	N_{21}	N_{22}
Subtotal	$N_{.1}$	$N_{.2}$

Table 2.6 Illustration of the BH (Benjamini and Hochberg 1995) and BY (Benjamini and Yekutieli 2001) procedures in controlling for FDR, with 15 sorted p values taken from Benjamini and Hochberg (1995)

i	p	P$_{critical\ .\ BH.i}$	P$_{critical\ .\ BY\ .\ i}$
1	0.0001	0.00333	0.00100
2	0.0004	0.00667	0.00201
3	0.0019	0.01000	0.00301
4	0.0095	0.01333	0.00402
5	0.0201	0.01667	0.00502
6	0.0278	0.02000	0.00603
7	0.0298	0.02333	0.00703
8	0.0344	0.02667	0.00804
9	0.0459	0.03000	0.00904
10	0.324	0.03333	0.01005
11	0.4262	0.03667	0.01105
12	0.5719	0.04000	0.01205
13	0.6528	0.04333	0.01306
14	0.759	0.04667	0.01406
15	1	0.05000	0.01507

The classical FDR approach (Benjamini and Hochberg 1995), now commonly referred to as the BH procedure, computes $p_{critical\ .\ BH.i}$ for the ith p value (where the subscript BH stands for the BH procedure) as

$$P_{critical\ .\ BH\ .\ i} = \frac{q \bullet i}{m} \qquad (2.20)$$

where q is FDR (e.g., 0.05), and i is the rank of the p value in the sorted array of p values (Table 6). If k is the largest i satisfying the condition of $p_i \leq p_{critical\ .\ BH.i}$, then we reject hypotheses from H_1 to H_k. In Table 6, k = 4 and we reject the first four hypotheses. Note that the fourth hypothesis was not rejected by $p_{critical\ .\ Bonferroni}$ but rejected by $p_{critical\ .\ BH.4}$. Also note that $p_{critical\ .\ Bonferroni}$ is the same as $p_{critical\ .\ BH.1}$.

The FDR procedure above assumes that the test statistics are independent. A more conservative FDR procedure has been developed that relaxes the independence assumption (Benjamini and Yekutieli 2001). This method, now commonly referred to as the BY procedure, computes $p_{critical\ .\ BY\ .\ i}$ for the ith hypothesis as

$$P_{critical.BY.i} = \frac{q \bullet i}{m \sum_{i=1}^{m} \frac{1}{i}} = \frac{P_{critical.BH.i}}{\sum_{i=1}^{m} \frac{1}{i}} \qquad (2.21)$$

With m = 15 in our case, Σ1/k = 3.318228993. Now k (the largest i satisfying $p_i \leq p_{critical\ .\ BY\ .\ i}$) is 3 (Table 2.6). Thus, only the first three hypotheses are rejected. The BY procedure was found to be too conservative and several alternatives have been proposed (Ge et al. 2008). For large m, $\Sigma 1/k$ converges to $\ln(m) + \gamma$ (Euler's constant equal approximately to 0.57721566). Thus, for $m = 10000$, $\Sigma 1/k$ is close to 10. So $p_{critical \cdot BY}$ is nearly 10 times smaller than $p_{critical \cdot BH}$.

One may also obtain empirical distribution of p values by resampling the data. For studying association between genes or between gene and environmental factors, one may compute the frequencies of states 0 (absence) and 1 (presence) for each gene (designated f_0 and f_1, respectively) and reconstitute each column by randomly sampling from the pool of states with f_0 and f_1. For each resampling, we may carry out the likelihood ratio test shown above to obtain p values. If we have generated 10000 p values, then the 500th smallest p value may be taken as the critical p value. Note that all the null hypotheses from resampled data are true. So FDR and FWER are equivalent. This is easy to see given that FDR is defined as the expected proportion of $N_{12}/N_{.2}$ (Table 2.5) and FWER as the probability that N_{12} (Table 2.5) is greater or equal to 1. As we cannot observe N_{ij}, we use n_{ij} to indicate their realized values. When all null hypotheses are true, $n_{22} = 0$ and $n_{12} = n_{.2}$. Now if $n_{12} > 0$, then FDR $= E(n_{12}/n_{.2}) = 1$, and FWER $= P(n_{12} \geq 1)$ is naturally also 1. If $n_{12} = 0$, then FDR $= 0$ (Recall that FDR is defined to be 0 when $n_{.2} = 0$), and FWER $= P(n_{12} \geq 1)$ is also 0 (Benjamini and Hochberg 1995).

Key steps in comparative genomics involves the compilation of data in the form of Fig. 2.1, perform data analyses such as measuring association and assessing statistical significance, and present biologically significant results. All these can be computationally automated with various software packages such as DAMBE which was originally released in 2000 (Xia 2001; Xia and Xie 2001). A recently developed XML-based rich data format, named NeXML (Vos et al. 2012), is particularly suitable for large-scale comparative genomics. Reconstruction of phylogenetic relationships from a large number of species and dating their branching points can be accomplished very rapidly by using the distance-based least-square methods (Xia and Yang 2011).

Postscript

Comparative genomic analysis depends on how we represent a genome. In Fig. 2.1, we represented a genome simply by the presence/absence of a set of genes, without paying any attention to gene locations on the genome. Although the function of most genes is independent of their genomic location, there are cases where positional information can help us in many scenarios of evolutionary inference. In most other books on comparative genomics, the genome is represented as a linear permutation of genes as signed or unsigned numbers, with the length and nature of the intergenic sequences ignored. This representation is used most frequently in studying genome rearrangement (Bader et al. 2001; Berman et al. 2002; Kececioglu and Sankoff 1994, 1995). The similarity between two genomes is often measured by the inversion distance (the minimum number of inversions that can convert one genome into another in gene order). If the inversion distance is significantly shorter than random expectation (which can be assessed by comparing the inversion distance against those from random permutation), we may assume that the two genomes share conserved synteny which simply means the similarity in the connectedness of a shared set of genes between two genomes. Here I briefly mention a few cases where conserved synteny provides invaluable information in comparative genomics.

Suppose that a processed pseudogene is found in two sister lineages, designated P_1 and P_2. Do P_1 and P_2 originate independently or inherited from their common ancestor? Because it is unlikely that two independent origins would land the pseudogene at the same genomic location in two separate lineages, we may answer the question by examining synteny conservation. Conserved synteny between the two lineages (Fig. 2.10a) indicates duplication before the split of the two lineages, whereas a lack of synteny (Fig. 2.10b) suggests that P_1 and P_2 may have originated independently, after the split of the two lineages.

Conserved synteny can also aid the identification of orthologous genes. Suppose two duplicated genes (P_1 and P_2) are found in two different genome segments (Fig. 2.10b) in species A. Also suppose that we have a gene P_3 in species B that matches both (P_1 and P_2). Is P_3 orthologous to P_1 or to P_2? Again, if P_3 shares synteny with P_1 but not with P_2, then we can conclude that P_3 and P_1 are orthologous.

It has often been assumed to be true that two rounds of genome duplication have occurred during vertebrate genome evolution, one in the ancestral chordate lineage and another after the divergence of gnathostomes from Agnatha. One more round of genome duplication has also been assumed to have occurred in the lineage leading to Salmonids. Genome duplication should lead to extensive synteny among chromosome segments. However, the limited synteny observed, coupled with a lack of quantitative assessment of the available evidence, proved to be insufficient to distinguish the hypothesis of genome duplication from the alternative hypothesis of duplication of genes and multi-gene segments (Hughes 2000).

Another way to represent a genome is to list the palindromes along the sequence, with each palindrome characterized by the length of the stem (number of paired nucleotides) and the length of the loop. Such a list of palindromes

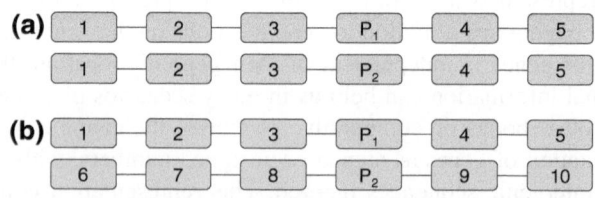

Fig. 2.10 Synteny helps us infer the timing of gene duplication and establish gene orthology. The boxes sharing the same number represent homologous genes **a** synteny present **b** synteny absent

constitutes the most fundamental aspect of genome phenotype or DNA morphology. Palindromes are the signposts of many cellular proteins involved in gene regulation, and a change in palindrome structure is expected to have functional implications.

Comparative genomics is a developing field. Better description of the genome will lead to new ways of representing the genome and new way of analysis. This chapter represents only a starting point in studying comparative genomics.

Chapter 3
Comparative Viral Genomics: Detecting Recombination

There are two major reasons to study recombination. The first is that it is biologically interesting. For example, different strains of viruses often recombine to form new strains of recombinants leading to host-jumping or resistance to antiviral medicine, posing direct threat to our health. The second reason is that recombination is the source of many evils in comparative genomics and molecular evolution as it can generate rate variation among sites and among lineages and distort phylogenetic relationships (Lemey and Posada 2009). We may be led astray without controlling for the effect of recombination in comparative genomic analysis.

Detecting viral recombination and mapping recombination points represent important research themes in viral comparative genomics (Salminen and Martin 2009). This is often done in two different situations. The first is to address whether one particular genome (typically the one causing human health concerns, designated hereafter as R) is the result of viral recombination from a set of N potential parental strains (designated hereafter as P_i, where i = 1, 2, ..., N). Graphic visualization methods such as Simplot (Lole et al. 1999) and Bootscan (Salminen et al. 1995), as well as the phylogenetic incongruence test (illustrated in Chap. 2), are often used in this first situation.

In the second situation, one does not know which one is R and which ones are P genomes. One simply has a set of genomic sequences and wishes to know whether some are recombinants of others. This is a more difficult problem. Many methods have been developed to solve the problem, and have been reviewed lucidly (Husmeier and Wright 2005). I will include here only what has been left out in the review, i.e., the graphic methods (Simplot and Bootscan) for the first situation and the compatibility matrix methods for the second. The compatibility matrix methods are among the most powerful methods for detecting recombination events.

Is a Particular Genome a Recombinant of N Other Genomes?

Given a sequence alignment, compute genetic distances $d_{R,Pi}$ (between R and P_i) along a sliding window of typically a few hundred bases. If we have a small $d_{R,Pi}$ and a large $d_{R,Pk}$ for one stretch of the genome, but a large $d_{R,Pi}$ and a small $d_{R,Pk}$ for another stretch of the genome, then a recombination likely occurred. This method, with visualization of the d values along the sliding windows, is known as Simplot (Lole et al. 1999). Its disadvantage is that it does not generate any measure of statistical confidence.

I will illustrate the Simplot procedure by using HIV-1 M genomes in an A-J-cons-kal153.fsa file (Salminen and Martin 2009). HIV-1 has three groups designated M (main), O (outgroup) and N (non-M and non-O), with the M group further divided into A-D and F-K subtypes. The A-J-cons-kal153.fsa contains consensus genomic sequences for subtypes A, B, C, D, F, G, H, and J, as well as the KAL153 strain which may be a recombinant of two of the subtypes.

The result of applying the Simplot procedure is shown in Fig. 3.1. The genetic distance used is a simultaneously estimated (SE) distance based on the F84 model (Xia 2009). Note that $d_{KAL153,A}$ is relatively small and $d_{KAL153,B}$ relatively large up to site 2601, after which $d_{KAL153,A}$ becomes large and $d_{KAL153,B}$ small until site 8701. After site 8701, $d_{KAL153,A}$ again becomes small and $d_{KAL153,B}$ large

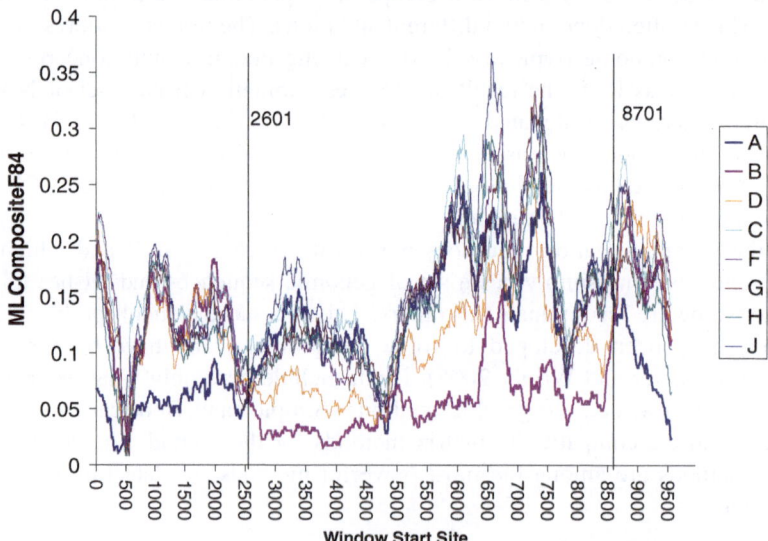

Fig. 3.1 Genetic distance between the query sequence (KAL153) and the consensus subtype sequences (A to J). MLCompositeF84 (Xia 2009) is a simultaneously estimated distance based on the F84 model. KAL153 is genetically close to A before window start site at 2601 and after window start site 8701, but becomes close to B between window start sites 2601 and 8701. Output from DAMBE (Xia 2001; Xia and Xie 2001)

(Fig. 3.1). The simplest interpretation is that KAL153 is a recombinant between an A-like strain and a B-like strain. The two sites at which KAL153 changes its phylogenetic affinity (i.e., 2601 and 8701) may be taken as the recombination sites.

One may ask what the interpretation would be if B is missing from the data. The interpretation unavoidably would be that KAL153 is a recombinant between an A-like strain and a D-like strain (Fig. 3.1). This interpretation is still reasonable because subtypes B and D are the most closely related phylogenetically. However, if A is missing from the data set, then the recombination event would become difficult to identify.

One might also note a few locations where the HIV-1 viral genomes are highly conserved across all included subtypes. Biopharmaceutical researchers typically would use such comparative genomic method to find conserved regions as drug targets or for developing vaccines against the virus.

One shortcoming of the Simplot method is that it does not produce any measure of statistical confidence. Given the stochastic nature of evolution, the distance of a sequence to other homologous sequence will often fluctuate. So the interpretation of patterns in Fig. 3.1 is associated with much uncertainty. Two approaches have been developed to overcome this shortcoming, one being the Bootscan method (Salminen and Martin 2009; Salminen et al. 1995), and the other is the phylogenetic incongruence test illustrated in Chap. 2.

The Bootscan method also takes a sliding window approach, but bootstraps the sequences to find the number of times each P_i has the smallest distance to R. The application of the bootscan method to the HIV-1 M data (Fig. 3.2) shows that A

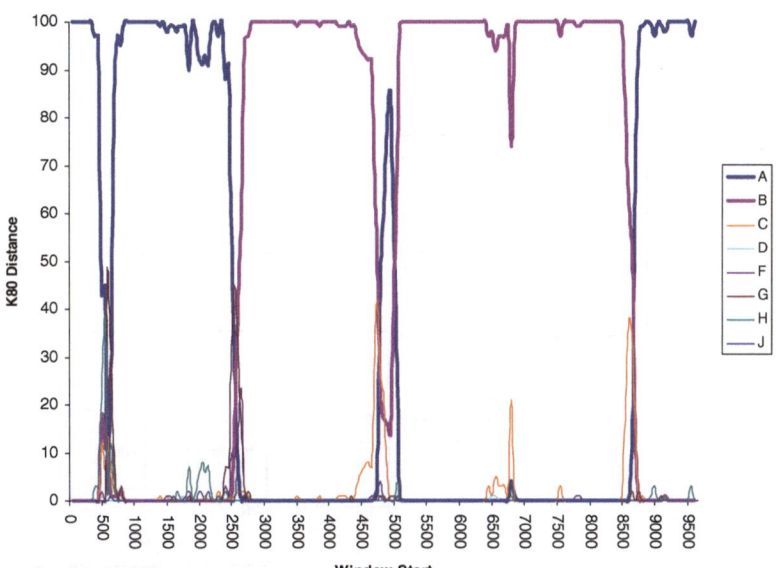

Fig. 3.2 BootScan output from scanning the HIV-1 M sequences with KAL153 as the query. Output from DAMBE (Xia 2001; Xia and Xie 2001), with window size being 400 nt and step size being 50 nt

is closest to KAL153 for almost all resampled data up to site 2601, after which B becomes the closest to KAL153 until site 4801. At this point A again becomes the closest to KAL153, albeit only briefly and with limited support. After site 5051, B again becomes the closest to KAL153 until site 8701 after which A again becomes the closest to KAL153 (Fig. 3.2). The result suggests that there might be two recombination events.

The Simplot and the Bootscan procedures work well with highly diverged parental sequences, e.g., when the parental sequences belong to different subtypes as in our examples above. Their performance is improved when genetic distances based on more realistic substitution models than K80 are used. DAMBE (Xia 2001; Xia and Xie 2001) implements many other distances including the GTR distance and several simultaneously estimated distances suitable for highly diverged sequences. However, the two methods are not sensitive when the parental sequences are closely related. This is true for most of the conventional methods for detecting recombination.

The second method for confirming KAL153s phylogenetic affinity reflected by changes in the genetic distance to other HIV-1 M genomes (Fig. 3.1) is the phylogenetic incongruence test. The result in Fig. 3.1 allows us to partition the aligned genomic sequences into two sets, one consisting of the segment from 2601 and 8630 (hereafter referred to MIDDLE), and the other made of the rest of the sequences (hereafter referred to as TAILS). The phylogenetic tree for the eight subtypes of HIV-1 M is shown in Fig. 3.3. A new HIV-1 M genome suspected to

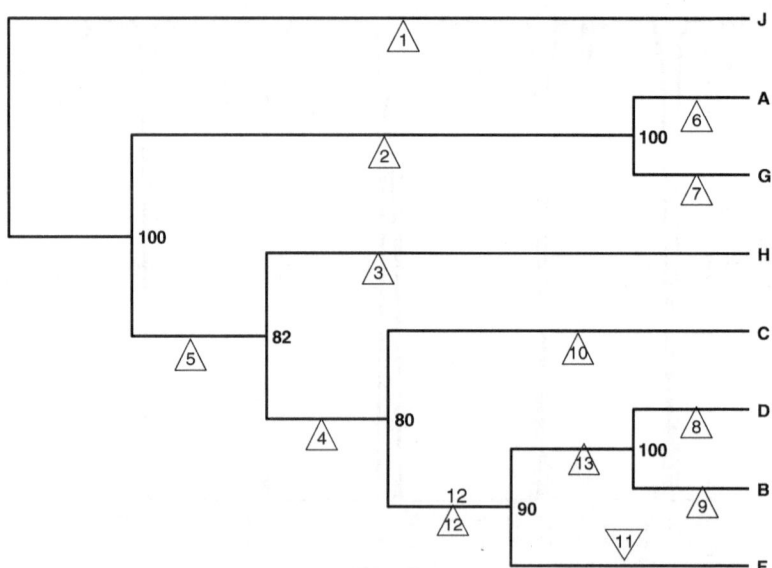

Fig. 3.3 Phylogenetic tree of the eight HIV-1 M subtype genomes, with percentage bootstrap support indicated at each internal node. The numbered arrows indicate branches to which KAL153 can be grafted to generate a new tree

be a recombinant, such as KAL153, may be phylogenetically grafted onto any one of the positions indicated by the numbered arrows (Fig. 3.3), creating 13 possible unrooted trees referred hereafter as T_1, T_2, ..., T_{13}, respectively, with the subscript number corresponding to the numbers in the arrow in Fig. 3.3). From results in Fig. 3.1, we can already infer that T_6 should be supported by the TAILS data set and T_9 should be supported the MIDDLE data set. However, will the support be significant against other alternative trees?

The result of phylogenetic tests (Table 3.1) shows that the TAILS data set strongly support T_6 (grouping KAL153 with subtype A) but the MIDDLE data set strongly support T_9 (grouping KAL153 with subtype B). This suggests that KAL153 is very highly likely to be a recombinant from subtypes A and B.

Table 3.1 Statistical tests of 13 alternative trees, based on the TAILS and MIDDLE data sets

Data	Tree	lnL[a]	ΔlnL[b]	SE (Δ)[c]	T	pT[d]	pSH[e]	pRELL[f]
TAILS	6	−15046.0	0.000	0.000				1.000
	2	−15223.6	−177.587	28.579	6.214	0.000	0.000	0.000
	7	−15225.4	−179.382	28.092	6.385	0.000	0.000	0.000
	1	−15279.4	−233.325	34.684	6.727	0.000	0.000	0.000
	5	−15287.2	−241.162	34.013	7.090	0.000	0.000	0.000
	3	−15334.1	−288.028	38.281	7.524	0.000	0.000	0.000
	4	−15341.0	−294.930	38.052	7.751	0.000	0.000	0.000
	10	−15373.2	−327.121	40.059	8.166	0.000	0.000	0.000
	12	−15379.0	−332.934	39.987	8.326	0.000	0.000	0.000
	11	−15423.2	−377.209	42.205	8.938	0.000	0.000	0.000
	13	−15424.7	−378.629	41.968	9.022	0.000	0.000	0.000
	9	−15592.2	−546.125	48.274	11.313	0.000	0.000	0.000
	8	−15598.1	−552.052	47.741	11.563	0.000	0.000	0.000
MIDDLE	9	−23875.2	0.000	0.000				1.000
	13	−24086.1	−210.934	30.721	6.866	0.000	0.000	0.000
	8	−24091.5	−216.388	30.005	7.212	0.000	0.000	0.000
	11	−24395.1	−519.977	48.357	10.753	0.000	0.000	0.000
	12	−24398.1	−522.909	47.870	10.924	0.000	0.000	0.000
	10	−24535.3	−660.101	54.873	12.030	0.000	0.000	0.000
	4	−24553.5	−678.299	54.061	12.547	0.000	0.000	0.000
	3	−24623.9	−748.766	56.714	13.202	0.000	0.000	0.000
	5	−24627.3	−752.148	56.671	13.272	0.000	0.000	0.000
	1	−24652.2	−776.994	57.503	13.512	0.000	0.000	0.000
	2	−24653.3	−778.099	57.767	13.470	0.000	0.000	0.000
	7	−24749.9	−874.732	61.169	14.300	0.000	0.000	0.000
	6	−24753.4	−878.281	61.246	14.340	0.000	0.000	0.000

[a] log-likelihood of each tree
[b] differences in log-likelihood between tree i and the best tree
[c] standard error of ΔlnL
[d] P value for paired-sample t test (two-tailed)
[e] P value with multiple-comparison correction (Shimodaira and Hasegawa 1999)
[f] RELL bootstrap proportions (Kishino and Hasegawa 1989)

The use of the MIDDLE and TAILS for the phylogenetic incongruence test might be criticized for having fallen into a sequential testing trap (Suchard et al. 2002). A sliding-window approach together with the control for the false discovery rate may be statistically more defendable.

General Methods Based on the Compatibility Matrix

In the set of four sequences in Fig. 3.4a, there are three possible unrooted trees labeled T_1, T_2 and T_3. Except for site 49, all sites are compatible with each other because they all support T_1. In contrast, site 49 supports T_3. In the classical population genetics with the infinite alleles model (Kimura and Crow 1964) where each mutation is unique and not reversible, site 49 would be considered as resulting from recombination because mutations, being unique and not reversible by definition with the infinite alleles model, could not produce the pattern in site 49. In other words, parallel convergent mutations in different evolutionary lineages (homoplasies) are not allowed in the infinite allele model.

The infinite alleles model is not applicable to nucleotide sequences where each site has only four possible states that can all change into each other. So we need to decide whether site 49 in Fig. 3.4a can be generated by substitutions without involving recombination. In general, sequence-based statistical methods for detecting recombination share one fundamental assumption (or flaw) that we have only

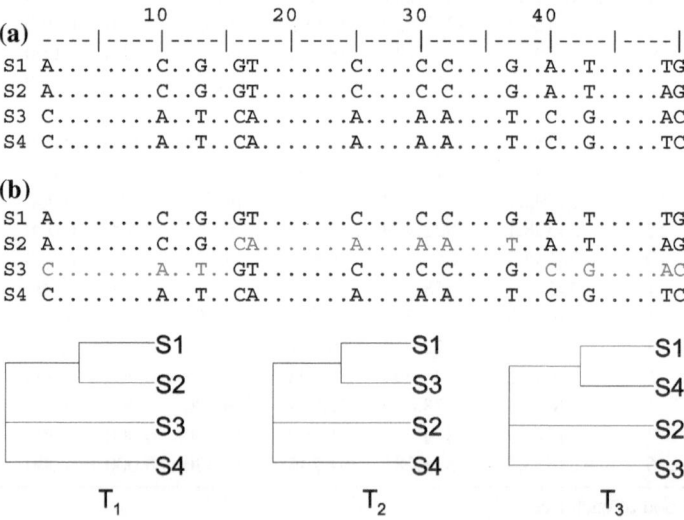

Fig. 3.4 Two sets of aligned nucleotide sequences for illustrating the compatibility-based method for detecting recombination events. **a** Four sequences without recombination. **b** Four sequences with recombination between S2 and S3, indicated by the switching of colored nucleotides. Dots indicate monomorphic sites

two alternatives, homoplasy or recombination, to explain polymorphic site patterns in a set of aligned sequences. If we reject the homoplasy explanation, then we arrive at the conclusion of recombination which is aptly named a backdoor conclusion (Hey 2000). Such a backdoor conclusion is ultimately not as satisfying as empirical demonstrations of recombination. For example, statistical detection of recombination involving mammalian mitochondrial genomes have been reported numerous times, but only an empirical demonstration (Kraytsberg et al. 2004) convinced the skeptical majority.

If we are happy with the fundamental assumption above that we have only two alternatives to discriminate between, then the method based on a compatibility matrix is both powerful and computationally fast. With a set of aligned sequences, two sites are compatible if and only if they both support the same tree topology. We only need to consider informative sites, i.e., sites featuring at least two states each of which is represented by at least two sequences. Non-informative sites are always compatible with other sites and need not be considered.

A pairwise compatibility matrix, or just compatibility matrix for short, lists whether sites i and j are compatible. The compatibility matrices for the two set of sequences in Fig. 3.4, one experiencing no recombination (Fig. 3.4a) and the other experiencing recombination involving the segment between informative sites 16–39 (Fig. 3.4b) are shown in Table 3.2. Two points are worth highlighting. First, sites that share the same evolutionary history are expected to be more compatible than those that do not (e.g., when the shared ancestry is disrupted by recombination). Note more 0's (compatible sites) in the upper triangle for sequences without recombination than in the lower triangle for sequences with recombination involving informative sites 16–39 (Table 3.2). Second, recombination tends to create similar neighbors in the compatibility matrix. Note the blocks of 1's and 0's in the lower triangle in Table 3.2. This similarity among neighbors has

Table 3.2 Pairwise compatibility matrices, with 0 for compatible sites and 1 for incompatible sites, for aligned sequences in Fig. 3.4a (*upper triangle*) without recombination and those in Fig. 3.4b (*lower triangle*) with recombination between informative sites 16–39

Site	1	10	13	16	17	25	30	32	37	40	43	49	50
1		0	0	0	0	0	0	0	0	0	0	1	0
10	0		0	0	0	0	0	0	0	0	0	1	0
13	0	0		0	0	0	0	0	0	0	0	1	0
16	1	1	1		0	0	0	0	0	0	0	1	0
17	1	1	1	0		0	0	0	0	0	0	1	0
25	1	1	1	0	0		0	0	0	0	0	1	0
30	1	1	1	0	0	0		0	0	0	0	1	0
32	1	1	1	0	0	0	0		0	0	0	1	0
37	1	1	1	0	0	0	0	0		0	0	1	0
40	0	0	0	1	1	1	1	1	1		0	1	0
43	0	0	0	1	1	1	1	1	1	0		1	0
49	1	1	1	1	1	1	1	1	1	1	1		1
50	0	0	0	1	1	1	1	1	1	0	0	1	

been characterized by the neighbor similarity score (NSS) which is the fraction of neighbors sharing either 0 (compatible) or 1 (incompatible). NSS is the basis of a number of methods for detecting recombination events (Brown et al. 2001; Jakobsen and Easteal 1996; Posada 2002; Posada and Crandall 2001; Wiuf et al. 2001) because its significance can be easily assessed by reshuffling the sites and recomputing NSS many times. The clumping of the compatible and incompatible sites in the compatibility matrix also suggests the possibility of mapping the recombination points. For example, one may infer from the compatibility matrix for the four sequences in Fig. 3.4b (lower triangle in Table 3.2) that the 5'-end recombination point is between informative sites 13 and 16, and that the 3'-end recombination point is between informative sites 37 and 40.

The compatibility matrix approach can be refined in two ways. First, when sequences are many, one will have some sites that are highly incompatible with each other as well as some sites that are only slightly incompatible with each other. The compatibility matrix approach lumps all these sites as incompatible sites, resulting in loss of information. Second, neighboring sites in a set of aligned sequences are expected to be more compatible with each other than with sites that are far apart. These two refinements were included in a recent study (Bruen et al. 2006) that uses a refined incompatibility score (RIS) and the PHI statistic based on RIS. This new method appears much more sensitive than previous ones based on empirical applications (Bruen et al. 2006; Salemi et al. 2008).

References

Akashi H, Gojobori T (2002) Metabolic efficiency and amino acid composition in the proteomes of *Escherichia coli* and *Bacillus subtilis*. Proc Natl Acad Sci U S A 99:3695–3700

Allen A, Flemstrom G, Garner A, Kivilaakso E (1993) Gastroduodenal mucosal protection. Physiol Rev 73:823–857

Altschul SF, Gish W, Miller W, Myers EW, Lipman DJ (1990) Basic local alignment search tool. J Mol Biol 215:403–410

Altschul SF, Madden TL, Schaffer AA, Zhang J, Zhang Z, Miller W, Lipman DJ (1997) Gapped BLAST and PSI-BLAST: a new generation of protein database search programs. Nucleic Acids Res 25:3389–3402

Argos P, Rossmann MG, Grau UM, Zuber A, Franck G, Tratschin JD (1979) Thermal stability and protein structure. Biochemistry (Mosc) 18:5698–5703

Aris-Brosou S, Xia X (2008) Phylogenetic analyses: a toolbox expanding towards Bayesian methods. Int J Plant Genomics 2008:683509

Bader DA, Moret BM, Yan M (2001) A linear-time algorithm for computing inversion distance between signed permutations with an experimental study. J Comput Biol 8:483–491

Bahir I, Fromer M, Prat Y, Linial M (2009) Viral adaptation to host: a proteome-based analysis of codon usage and amino acid preferences. Mol Syst Biol 5:311

Baik SC, Kim KM, Song SM, Kim DS, Jun JS, Lee SG, Song JY, Park JU, Kang HL, Lee WK, Cho MJ, Youn HS, Ko GH, Rhee KH (2004) Proteomic analysis of the sarcosine-insoluble outer membrane fraction of *Helicobacter pylori* strain 26695. J Bacteriol 186:949–955

Ballester R, Marchuk D, Boguski M, Saulino A, Letcher R, Wigler M, Collins F (1990) The NF1 locus encodes a protein functionally related to mammalian GAP and yeast IRA proteins. Cell 63:851–859

Barker D, Meade A, Pagel M (2007) Constrained models of evolution lead to improved prediction of functional linkage from correlated gain and loss of genes. Bioinformatics 23:14–20

Barker D, Pagel M (2005) Predicting functional gene links from phylogenetic-statistical analyses of whole genomes. PLoS Comput Biol 1:e3

Bauerfeind P, Garner R, Dunn BE, Mobley HL (1997) Synthesis and activity of *Helicobacter pylori* urease and catalase at low pH. Gut 40:25–30

Baumgartner HK, Montrose MH (2004) Regulated alkali secretion acts in tandem with unstirred layers to regulate mouse gastric surface pH. Gastroenterology 126:774–783

Benjamini Y, Hochberg Y (1995) Controlling the false discovery rate: a practical and powerful approach to multiple testing. J R Statist Soc B 57:289–300

Benjamini Y, Yekutieli D (2001) The control of the false discovery rate in multiple hypothesis testing under dependency. Ann Stat 29:1165–1188

Berman P, Hannenhalli S, Karpinski M (2002) 1.375-approximation algorithm for sorting by reversals. Eur Symp Algorithms 2461:401–408

Bestor TH, Coxon A (1993) The pros and cons of DNA methylation. Curr Biol 6:384–386

Bogenhagen DF, Clayton DA (2003) The mitochondrial DNA replication bubble has not burst. Trends Biochem Sci 28:357–360

Brown CJ, Garner EC, Keith Dunker A, Joyce P (2001) The power to detect recombination using the coalescent. Mol Biol Evol 18:1421–1424

Brown TA, Cecconi C, Tkachuk AN, Bustamante C, Clayton DA (2005) Replication of mitochondrial DNA occurs by strand displacement with alternative light-strand origins, not via a strand-coupled mechanism. Genes Dev 19:2466–2476

Bruen TC, Philippe H, Bryant D (2006) A simple and robust statistical test for detecting the presence of recombination. Genetics 172:2665–2681

Bulmer M (1987) Coevolution of codon usage and transfer RNA abundance. Nature 325:728–730

Bulmer M (1991) The selection-mutation-drift theory of synonymous codon usage. Genetics 129:897–907

Burge C, Karlin S (1997) Prediction of complete gene structures in human genomic DNA. J Mol Biol 268:78–94

Burge CB, Karlin S (1998) Finding the genes in genomic DNA. Curr Opin Struct Biol 8:346–354

Bury-Mone S, Skouloubris S, Labigne A, De Reuse H (2001) The *Helicobacter pylori* UreI protein: role in adaptation to acidity and identification of residues essential for its activity and for acid activation. Mol Microbiol 42:1021–1034

Cardon LR, Burge C, Clayton DA, Karlin S (1994) Pervasive CpG suppression in animal mitochondrial genomes. Proc Natl Acad Sci U.S.A 91:3799–3803

Carullo M, Xia X (2008) An extensive study of mutation and selection on the wobble nucleotide in tRNA anticodons in fungal mitochondrial genomes. J Mol Evol 66:484–493

Chambaud I, Heilig R, Ferris S, Barbe V, Samson D, Galisson F, Moszer I, Dybvig K, Wroblewski H, Viari A, Rocha EP, Blanchard A (2001) The complete genome sequence of the murine respiratory pathogen *Mycoplasma pulmonis*. Nucleic Acids Res 29:2145–2153

Chan PP, Lowe TM (2009) GtRNAdb: a database of transfer RNA genes detected in genomic sequence. Nucl Acids Res 37:D93–97

Clayton DA (1982) Replication of animal mitochondrial DNA. Cell 28:693–705

Clayton DA (2000) Transcription and replication of mitochondrial DNA. Hum Reprod 15:11–17

Colby C, Edlin G (1970) Nucleotide pool levels in growing, inhibited, and transformed chick fibroblast cells. Biochemistry (Mosc) 9:917

Crick FH (1966) Codon–anticodon pairing: the wobble hypothesis. J Mol Biol 19:548–555

Curran JF (1995) Decoding with the A:I wobble pair is inefficient. Nucleic Acids Res 23:683–688

Dalgaard JZ, Garrett RA (1993) Archaeal hyperthermophile genes. In: Kates M, Kushner DJ, Matheson AT (eds) The biochemistry of Archaea (*Archaebacteria*). Elsevier, Amsterdam

Drummond A, Rambaut A (2007) BEAST: Bayesian evolutionary analysis by sampling trees. BMC Evol Biol 7:214

Engel E, Peskoff A, Kauffman GL Jr, Grossman MI (1984) Analysis of hydrogen ion concentration in the gastric gel mucus layer. Am J Physiol 247:G321–G338

Felsenstein J (1981) Evolutionary trees from DNA sequences: a maximum likelihood approach. J Mol Evol 17:368–376

Felsenstein J (1985) Phylogenies and the comparative method. Am Nat 125:1–15

Felsenstein J (2002) PHYLIP 3.6 (phylogeny inference package). Version 3.6. Department of Genetics, University of Washington, Seattle

Felsenstein J (2004) Inferring phylogenies. Sinauer, Sunderland

Frederico LA, Kunkel TA, Shaw BR (1990) A sensitive genetic assay for the detection of cytosine deamination: determination of rate constants and the activation energy. Biochemistry (Mosc) 29:2532–2537

Frederico LA, Kunkel TA, Shaw BR (1993) Cytosine deamination in mismatched base pairs. Biochemistry (Mosc) 32:6523–6530

Galtier N, Lobry JR (1997) Relationships between genomic G+C content, RNA secondary structures, and optimal growth temperature in prokaryotes. J Mol Evol 44:632–636

Gardy JL, Spencer C, Wang K, Ester M, Tusnady GE, Simon I, Hua S, deFays K, Lambert C, Nakai K, Brinkman FS (2003) PSORT-B: Improving protein subcellular localization prediction for Gram-negative bacteria. Nucleic Acids Res 31:3613–3617

Ge Y, Sealfon SC, Speed TP (2008) Some step-down procedures controlling the false discovery rate under dependence. Stat Sin 18:881–904

Gojobori T, Li WH, Graur D (1982) Patterns of nucleotide substitution in pseudogenes and functional genes. J Mol Evol 18:360–369

Goto M, Washio T, Tomita M (2000) Causal analysis of CpG suppression in the *Mycoplasma* genome. Microb Comp Genomics 5:51–58

Gouy M, Gautier C (1982) Codon usage in bacteria: correlation with gene expressivity. Nucleic Acids Res 10:7055–7064

Grosjean H, de Crecy-Lagard V, Marck C (2010) Deciphering synonymous codons in the three domains of life: co-evolution with specific tRNA modification enzymes. FEBS Lett 584:252–264

Haas J, Park E-C, Seed B (1996) Codon usage limitation in the expression of HIV-1 envelope glycoprotein. Curr Biol 6:315–324

Hamajima N, Goto Y, Nishio K, Tanaka D, Kawai S, Sakakibara H, Kondo T (2004) *Helicobacter pylori* eradication as a preventive tool against gastric cancer. Asian Pac J Cancer Prev 5:246–252

Hanada K, Suzuki Y, Gojobori T (2004) A large variation in the rates of synonymous substitution for RNA viruses and its relationship to a diversity of viral infection and transmission modes. Mol Biol Evol 21:1074–1080

Harvey PH, Pagel MD (1991) The comparative method in evolutionary biology. Oxford University Press, Oxford

Hey J (2000) Human mitochondrial DNA recombination: can it be true? Trends Ecol Evol 15:181–182

Higgs PG, Ran W (2008) Coevolution of codon usage and tRNA genes leads to alternative stable states of biased codon usage. Mol Biol Evol 25:2279–2291

Hofer A, Steverding D, Chabes A, Brun R, Thelander L (2001) *Trypanosoma brucei* CTP synthetase: a target for the treatment of African sleeping sickness. Proc Natl Acad Sci U S A 98:6412–6416

Hughes AL (2000) Polyploidization and vertebrate origins: a review of the evidence. In: Sankoff D, Nadeau JH (eds) Comparative genomics. Kluwer Academic Publishers, Netherland, pp 493–502

Hunt RH (2004) Will eradication of *Helicobacter pylori* infection influence the risk of gastric cancer? Am J Med 117:86S–91S

Hurst LD, Merchant AR (2001) High guanine-cytosine content is not an adaptation to high temperature: a comparative analysis amongst prokaryotes. Proc R Soc Lond B 268:493–497

Husmeier D, Wright F (2005) Detectign recombination in DNA sequence alignments. In: Husmeier D, Dybowski R, Roberts S (eds) Probabilistic modeling in bioinformatics and medical informatics. Springer, London, p 504

Ikemura T (1981a) Correlation between the abundance of *Escheriachia coli* transfer RNAs and the occurrence of the respective codons in its protein genes. J Mol Biol 146:1–21

Ikemura T (1981b) Correlation between the abundance of *Escherichia coli* transfer RNAs and the occurrence of the respective codons in its protein genes: a proposal for a synonymous codon choice that is optimal for the *E coli* translational system. J Mol Biol 151:389–409

Ikemura T (1982) Correlation between the abundance of yeast transfer RNAs and the occurrence of the respective codons in protein genes. Differences in synonymous codon choice patterns of yeast and Escherichia coli with reference to the abundance of isoaccepting transfer RNAs. J Mol Biol 158:573–597

Ikemura T (1992) Correlation between codon usage and tRNA content in microorganisms. In: Hatfield DL, Lee BJ, Pirtle RM (eds) Transfer RNA in protein synthesis. CRC Press, Boca Raton, pp 87–111

Irimia M, Penny D, Roy SW (2007) Coevolution of genomic intron number and splice sites. Trends Genet 23:321

Jacob F (1988) The statue within: an autobiography. Basic Books, Inc, New York

Jakobsen IB, Easteal S (1996) A program for calculating and displaying compatibility matrices as an aid in determining reticulate evolution in molecular sequences. Comput Appl Biosci 12:291–295

Jenkins GM, Holmes EC (2003) The extent of codon usage bias in human RNA viruses and its evolutionary origin. Virus Res 92:1–7

Jia W, Higgs PG (2008) Codon usage in mitochondrial genomes: distinguishing context-dependent mutation from translational selection. Mol Biol Evol 25:339–351

Josse J, Kaiser AD, Kornberg A (1961) Enzymatic synthesis of deoxyribonucleic acid VII. Frequencies of nearest neighbor base-sequences in deoxyribonucleic acid. J Biol Chem 236:864–875

Karlin S, Burge C (1995) Dinucleotide relative abundance extremes: a genomic signature. TIG 11:283–290

Karlin S, Mrazek J (1996) What drives codon choices in human genes. J Mol Biol 262:459–472

Keating CP, Hill MK, Hawkes DJ, Smyth RP, Isel C, Le SY, Palmenberg AC, Marshall JA, Marquet R, Nabel GJ, Mak J (2009) The A-rich RNA sequences of HIV-1 pol are important for the synthesis of viral cDNA. Nucleic Acids Res 37:945–956

Kececioglu JD, Sankoff D (eds) (1994) In: In Proceedings of the 5th symposium on combinatorial pattern matching

Kececioglu JD, Sankoff D (1995) Exact and approximation algorithms for sorting by reversal. Algorithmica 13:180–210

Kimura M, Crow JF (1964) The number of alleles that can be maintained in a finite population. Genetics 49:725–738

Kishino H, Hasegawa M (1989) Evaluation of the maximum likelihood estimate of the evolutionary tree topologies from DNA sequence data, and the branching order in *Hominoidea*. J Mol Evol 29:170–179

Kliman RM, Bernal CA (2005) Unusual usage of AGG and TTG codons in humans and their viruses. Gene 352:92

Kraytsberg Y, Schwartz M, Brown TA, Ebralidse K, Kunz WS, Clayton DA, Vissing J, Khrapko K (2004) Recombination of human mitochondrial DNA. Science 304:981

Kumar S, Nei M, Dudley J, Tamura K (2008) MEGA: a biologist-centric software for evolutionary analysis of DNA and protein sequences. Brief Bioinform 9:299–306

Kushiro A, Shimizu M, Tomita K-I (1987) Molecular cloning and sequence determination of the *tuf* gene coding for the elongation factor Tu of *Thermus thermophilus* HB8. Eur J Biochem 170:93–98

Kypr J, Mrazek JAN (1987) Unusual codon usage of HIV. Nature 327:20

Lemey P, Posada D (2009) Introduction to recombination detection. In: Lemey P, Salemi M, Vandamme AM (eds) The phylogenetic handbook. Cambridge University Press, Cambridge, pp 493–518

Li W-H (1983) Evolution of duplicate genes and pseudogenes. Sinauer, Sunderland

Li WH, Gojobori T, Nei M (1981) Pseudogenes as a paradigm of neutral evolution. Nature 292:237–239

Lindahl T (1993) Instability and decay of the primary structure of DNA. Nature 362:709–715

Lobry JR (1996) Asymmetric substitution patterns in the two DNA strands of bacteria. Mol Biol Evol 13:660–665

Lole KS, Bollinger RC, Paranjape RS, Gadkari D, Kulkarni SS, Novak NG, Ingersoll R, Sheppard HW, Ray SC (1999) Full-length human immunodeficiency virus type 1 genomes from subtype C-infected seroconverters in India, with evidence of intersubtype recombination. J Virol 73:152–160

Lopez P, Philippe H, Myllykallio H, Forterre P (1999) Identification of putative chromosomal origins of replication in *Archaea*. Mol Microbiol 32:883–886

Lough J, Jackson M, Morris R, Moyer R (2001) Bisulfite-induced cytosine deamination rates in *E. coli* SSB:DNA complexes. Mutat Res 478:191–197

Ma P, Xia X (2011) Factors affecting splicing strength of yeast genes. Comp Funct Genomics Article ID 212146:13

Marin A, Xia X (2008) GC skew in protein-coding genes between the leading and lagging strands in bacterial genomes: new substitution models incorporating strand bias. J Theor Biol 253:508–513

Marín A, Xia X (2008) GC skew in protein-coding genes between the leading and lagging strands in bacterial genomes: new substitution models incorporating strand bias. J Theor Biol 253:508–513

Martinez MA, Vartanian J-P, Simon W-H (1994) Hypermutagenesis of RNA Using human immunodeficiency virus type 1 reverse transcriptase and biased dNTP concentrations. Proc Natl Acad Sci U S A 91:11787–11791

Martins EP, Hansen TF (1997) Phylogenies and the comparative method: a general approach to incorporating phylogenetic information into the analysis of interspecific data. Am Nat 149:646–667

Matin A, Zychlinsky E, Keyhan M, Sachs G (1996) Capacity of *Helicobacter pylori* to generate ionic gradients at low pH is similar to that of bacteria which grow under strongly acidic conditions. Infect Immun 64:1434–1436

Menaker RJ, Sharaf AA, Jones NL (2004) *Helicobacter pylori* infection and gastric cancer: host, bug, environment, or all three? Curr Gastroenterol Rep 6:429–435

Mendz GL, Hazell SL (1996) The urea cycle of *Helicobacter pylori*. Microbiology 142:2959–2967

Mendz GL, Jimenez BM, Hazell SL, Gero AM, O'Sullivan WJ (1994) De novo synthesis of pyrimidine nucleotides by *Helicobacter pylori*. J Appl Bacteriol 77:1–8

Mine T, Muraoka H, Saika T, Kobayashi I (2005) Characteristics of a clinical isolate of urease-negative *Helicobacter pylori* and its ability to induce gastric ulcers in *Mongolian gerbils*. Helicobacter 10:125–131

Mobley HL, Hu LT, Foxal PA (1991) *Helicobacter pylori* urease: properties and role in pathogenesis. Scand J Gastroenterol Suppl 187:39–46

Mushegian AR, Koonin EV (1996) A minimal gene set for cellular life derived by comparison of complete bacterial genomes. Proc Natl Acad Sci U S A 93:10268–10273

Muto A, Osawa S (1987) The guanine and cytocine content of genomic DNA and bacterial evolution. Proc Natl Acad Sci USA 84:166–169

Nakai K, Horton P (1999) PSORT: a program for detecting sorting signals in proteins and predicting their subcellular localization. Trends Biochem Sci 24:34–36

Nakamura Y, Gojobori T, Ikemura T (2000) Codon usage tabulated from international DNA sequence databases: status for the year 2000. Nucleic Acids Res 28:292

Nakashima H, Fukuchi S, Nishikawa K (2003) Compositional changes in RNA, DNA and proteins for bacterial adaptation to higher and lower temperatures. J Biochem (Tokyo) 133:507–513

Ngumbela KC, Ryan KP, Sivamurthy R, Brockman MA, Gandhi RT, Bhardwaj N, Kavanagh DG (2008) Quantitative effect of suboptimal codon usage on translational efficiency of mRNA encoding HIV-1 *gag* in Intact T Cells. PLoS ONE 3:e2356

Nichols T, Hayasaka S (2003) Controlling the familywise error rate in functional neuroimaging: a comparative review. Stat Meth Med Res 12:419–446

Nur I, Szyf M, Razin A, Glaser G, Rottem S, Razin S (1985) Procaryotic and eucaryotic traits of DNA methylation in *Spiroplasmas* (*Mycoplasmas*). J Bacteriol 164:19–24

Nussinov R (1984) Doublet frequencies in evolutionary distinct groups. Nucleic Acids Res 12:1749–1763

Ochman H, Lawrence JG, Groisman EA (2000) Lateral gene transfer and the nature of bacterial innovation. Nature 405:299–304

Page RDM (2003) Introduction. In: Page RDM (ed) Tangled trees: phylogeny, cospeciation and coevolution. University of Chicago Press, Chicago, pp 1–21

Pagel M (1994) Detecting correlated evolution on phylogenies: a general method for the comparative analysis of discrete characters. Proc R Soc Lond B Biol Sci 255:37–45

Pagel M (1997) Inferring evolutionary processes from phylogenies. Zool Scr 26:331–348

Pagel M (1999) Inferring the historical patterns of biological evolution. Nature 401:877–884

Palidwor GA, Perkins TJ, Xia X (2010) A general model of codon bias due to GC mutational bias. PLoS One 5:e13431

Paterson AM, Gray RD, Wallis GP (1995) Of lice and men: the return of the 'comparative parasitology' debate. Parasitol Today 11:158–160

Plotkin JB, Kudla G (2010) Synonymous but not the same: the causes and consequences of codon bias. Nat Rev Genet 12:32–42

Plotkin JB, Robins H, Levine AJ (2004) Tissue-specific codon usage and the expression of human genes. Proc Natl Acad Sci U S A 101:12588–12591

Posada D (2002) Evaluation of methods for detecting recombination from DNA sequences: empirical data. Mol Biol Evol 19:708–717

Posada D, Crandall KA (2001) Evaluation of methods for detecting recombination from DNA sequences: computer simulations. Proc Natl Acad Sci U S A 98:13757–13762

Press WH, Teukolsky SA, Tetterling WT, Flannery BP (1992) Numerical recipes in C: the art of scientific computing. Cambridge University Press, Cambridge

Razin A, Razin S (1980) Methylated bases in mycoplasmal DNA. Nucleic Acids Res 8:1383–1390

Rektorschek M, Buhmann A, Weeks D, Schwan D, Bensch KW, Eskandari S, Scott D, Sachs G, Melchers K (2000) Acid resistance of *Helicobacter pylori* depends on the UreI membrane protein and an inner membrane proton barrier. Mol Microbiol 36:141–152

Rideout WMI, Coetzee GA, Olumi AF, Jones PA (1990) 5-Methylcytosine as an endogenous mutagen in the human LDL receptor and p53 genes. Science 249:1288–1290

Rimsky L, Hauber J, Dukovich M, Malim MH, Langlois A, Cullen BR, Greene WC (1988) Functional replacement of the HIV-1 rev protein by the HTLV-1 rex protein. Nature 335:738–740

Rocha EP, Danchin A (2002) Base composition bias might result from competition for metabolic resources. Trends Genet 18:291–294

Sachs G, Meyer-Rosberg K, Scott DR, Melchers K (1996) Acid, protons and *Helicobacter pylori*. Yale J Biol Med 69:301–316

Sachs G, Weeks DL, Melchers K, Scott DR (2003) The gastric biology of *Helicobacter pylori*. Annu Rev Physiol 65:349–369

Saenger W (1984) Principles of nucleic acid structure. Springer, New York

Salemi M, Gray RR, Goodenow MM (2008) An exploratory algorithm to identify intra-host recombinant viral sequences. Mol Phylogenet Evol 49:618

Salminen M, Martin D (2009) Detecting and characterizing individual recombination events. In: Lemey P, Salemi M, Vandamme AM (eds) The phylogenetic handbook. Cambridge University Press, Cambridge, pp 519–548

Salminen MO, Carr JK, Burke DS, McCutchan FE (1995) Identification of breakpoints in intergenotypic recombinants of HIV type 1 by bootscanning. AIDS Res Hum Retroviruses 11:1423–1425

Sancar A, Sancar GB (1988) DNA repair enzymes. Annu Rev Biochem 57:29–67

Sansonetti PJ, d'Hauteville H, Formal SB, Toucas M (1982a) Plasmid-mediated invasiveness of "Shigella-like" *Escherichia coli*. Ann Microbiol (Paris) 133:351–355

Sansonetti PJ, Kopecko DJ, Formal SB (1982b) Involvement of a plasmid in the invasive ability of *Shigella flexneri*. Infect Immun 35:852–860

Schluter D, Price TD, Mooers AØ, Ludwig D (1997) Likelihood of ancestor states in adaptive radiation. Evolution 51:1699–1711

Scott D, Weeks D, Melchers K, Sachs G (1998) The life and death of *Helicobacter pylori*. Gut 43:S56–60

Scott DR, Marcus EA, Weeks DL, Sachs G (2002) Mechanisms of acid resistance due to the urease system of *Helicobacter pylori*. Gastroenterology 123:187–195

Shadel GS, Clayton DA (1997) Mitochondrial DNA maintenance in vertebrates. Annu Rev Biochem 66:409–435

Sharp PM (1986) What can AIDS virus codon usage tell us? Nature 324:114

Sharp PM, Li WH (1987) The codon adaptation Index a measure of directional synonymous codon usage bias, and its potential applications. Nucleic Acids Res 15:1281–1295

Shimodaira H, Hasegawa M (1999) Multiple comparisons of log-likelihoods with applications to phylogenetic inference. Mol Biol Evol 16:1114–1116

Siavoshi F, Malekzadeh R, Daneshmand M, Smoot DT, Ashktorab H (2004) Association between *Helicobacter pylori* Infection in gastric cancer, ulcers and gastritis in Iranian patients. Helicobacter 9:470

Singer CE, Ames BN (1970) Sunlight ultraviolet and bacterial DNA base ratios. Science 170:822–826

Slonczewski JL, Rosen BP, Alger JR, Macnab RM (1981) pH homeostasis in *Escherichia coli*: measurement by 31P nuclear magnetic resonance of methylphosphonate and phosphate. Proc Natl Acad Sci U S A 78:6271–6275

Stingl K, Altendorf K, Bakker EP (2002a) Acid survival of *Helicobacter pylori*: how does urease activity trigger cytoplasmic pH homeostasis? Trends Microbiol 10:70–74

Stingl K, Uhlemann E-M, Schmid R, Altendorf K, Bakker EP (2002b) Energetics of *Helicobacter pylori* and its implications for the mechanism of urease-dependent acid tolerance at pH 1. J Bacteriol 184:3053–3060

Stingl K, Uhlemann EM, Deckers-Hebestreit G, Schmid R, Bakker EP, Altendorf K (2001) Prolonged survival and cytoplasmic pH homeostasis of *Helicobacter pylori* at pH 1. Infect Immun 69:1178–1180

Stoebel DM (2005) Lack of evidence for horizontal transfer of the lac operon into *Escherichia coli*. Mol Biol Evol 22:683–690

Strebel K (2005) APOBEC3G & HTLV-1: inhibition without deamination. Retrovirology 2:37

Suchard MA, Weiss RE, Dorman KS, Sinsheimer JS (2002) Oh brother, where art thou? A Bayes factor test for recombination with uncertain heritage. Syst Biol 51:715–728

Sueoka N (1964) On the evolution of informational macromolecules. Academic Press, New York

Sved J, Bird A (1990) The expected equilibrium of the CpG dinucleotide in vertebrate genomes under a mutation model. Proc Natl Acad Sci USA 87:4692–4696

Swofford DL (2000) Phylogeentic analysis using parsimony (* and other methods). Sinauer, Sunderland

Tanaka M, Ozawa T (1994) Strand asymmetry in human mitochondrial DNA mutations. Genomics 22:327–335

Tomb JF, White O, Kerlavage AR, Clayton RA, Sutton GG, Fleischmann RD, Ketchum KA, Klenk HP, Gill S, Dougherty BA, Nelson K, Quackenbush J, Zhou L, Kirkness EF, Peterson S, Loftus B, Richardson D, Dodson R, Khalak HG, Glodek A, McKenney K, Fitzegerald LM, Lee N, Adams MD, Venter JC et al (1997) The complete genome sequence of the gastric pathogen *Helicobacter pylori*. Nature 388:539–547

Valenzuela M, Cerda O, Toledo H (2003) Overview on chemotaxis and acid resistance in *Helicobacter pylori*. Biol Res 36:429–436

Van Dooren S, Pybus OG, Salemi M, Liu HF, Goubau P, Remondegui C, Talarmin A, Gotuzzo E, Alcantara LC, Galvao-Castro B, Vandamme AM (2004) The low evolutionary rate of human T-cell lymphotropic virus type-1 confirmed by analysis of vertical transmission chains. Mol Biol Evol 21:603–611

van Hemert FJ, Berkhout B (1995) The tendency of lentiviral open reading frames to become A-rich: constraints imposed by viral genome organization and cellular tRNA availability. J Mol Evol 41:132–140

van Weringh A, Ragonnet-Cronin M, Pranckeviciene E, Pavon-Eternod M, Kleiman L, Xia X (2011) HIV-1 modulates the tRNA pool to improve translation efficiency. Mol Biol Evol 28:1827–1834

Vartanian J-P, Henry M, Wain-Hobson S (2002) Sustained G→A hypermutation during reverse transcription of an entire human immunodeficiency virus type 1 strain Vau group O genome. J Gen Virol 83:801–805

Vinci G, Xia X, Veitia RA (2008) Preservation of genes involved in sterol metabolism in cholesterol auxotrophs: facts and hypotheses. PLoS ONE 3:e2883

Vos RA, Balhoff JP, Caravas JA, Holder MT, Lapp H, Maddison WP, Midford PE, Priyam A, Sukumaran J, Xia X, Stoltzfus A (2012) NeXML: rich, extensible, and verifiable representation of comparative data and metadata. Syst Biol 61:675–689

Wang HC, Hickey DA (2002) Evidence for strong selective constraint acting on the nucleotide composition of 16S ribosomal RNA genes. Nucleic Acids Res 30:2501–2507

Wang HC, Xia X, Hickey DA (2006) Thermal adaptation of ribosomal RNA genes: a comparative study. J Mol Evol 63:120–126

Weber MJ, Edlin G (1971) Phosphate transport, nucleotide pools, and ribonucleic acid synthesis in growing and in density-inhibited 3T3 cells. J Biol Chem 246:1828–1833

Weeks DL, Eskandari S, Scott DR, Sachs G (2000) A H+-gated urea channel: the link between *Helicobacter pylori* urease and gastric colonization. Science 287:482–485

Wen Y, Marcus EA, Matrubutham U, Gleeson MA, Scott DR, Sachs G (2003) Acid-adaptive genes of *Helicobacter pylori*. Infect Immun 71:5921–5939

Williams CL, Preston T, Hossack M, Slater C, McColl KE (1996) *Helicobacter pylori* utilises urea for amino acid synthesis. FEMS Immunol Med Microbiol 13:87–94

Wiuf C, Christensen T, Hein J (2001) A simulation study of the reliability of recombination detection methods. Mol Biol Evol 18:1929–1939

Wright F (1990) The 'effective number of codons' used in a gene. Gene 87:23–29

Xia X (1996) Maximizing transcription efficiency causes codon usage bias. Genetics 144:1309–1320

Xia X (1998a) How optimized is the translational machinery in *Escherichia coli*, *Salmonella typhimurium* and *Saccharomyces cerevisiae*? Genetics 149:37–44

Xia X (1998b) The rate heterogeneity of nonsynonymous substitutions in mammalian mitochondrial genes. Mol Biol Evol 15:336–344

Xia X (2001) Data analysis in molecular biology and evolution. Kluwer Academic Publishers, Boston

Xia X (2003) DNA methylation and mycoplasma genomes. J Mol Evol 57:S21–S28

Xia X (2005) Mutation and selection on the anticodon of tRNA genes in vertebrate mitochondrial genomes. Gene 345:13–20

Xia X (2007a) Bioinformatics and the cell: modern computational approaches in genomics, proteomics and transcriptomics. Springer, New York

Xia X (2007b) An improved implementation of codon adaptation index. Evol Bioinform 3:53–58

Xia X (2008) The cost of wobble translation in fungal mitochondrial genomes: integration of two traditional hypotheses. BMC Evol Biol 8:211

Xia X (2009) Information-theoretic indices and an approximate significance test for testing the molecular clock hypothesis with genetic distances. Mol Phylogenet Evol 52:665–676

Xia X (2012a) DNA replication and strand asymmetry in prokaryotic and mitochondrial genomes. Current Genomics 13:16–27

Xia X (2012b) Rapid evolution of animal mitochondria. In: Singh RS, Xu J, Kulathinal RJ (eds) Evolution in the fast lane: rapidly evolving genes and genetic systems. Oxford University Press, Oxford

Xia X (2013) Wobble hypothesis. In: Maloy S, Hughes K (eds) Brenner's encyclopedia of genetics. 2nd edition, Academic Press

Xia X, Huang H, Carullo M, Betran E, Moriyama EN (2007) Conflict between translation Initiation and elongation in vertebrate mitochondrial genomes. PLoS ONE 2:e227

Xia X, Li WH (1998) What amino acid properties affect protein evolution? J Mol Evol 47:557–564

Xia X, MacKay V, Yao X, Wu J, Miura F, Ito T, Morris DR (2011) Translation initiation: a regulatory role for Poly(A) tracts in front of the AUG codon in *Saccharomyces cerevisiae*. Genetics 189:469–478

Xia X, Palidwor G (2005) Genomic adaptation to acidic environment: evidence from *Helicobacter pylori*. Am Nat 166:776–784

Xia X, Wang HC, Xie Z, Carullo M, Huang H, Hickey DA (2006) Cytosine usage modulates the correlation between CDS length and CG content in prokaryotic genomes. Mol Biol Evol 23:1450–1454

Xia X, Xie Z (2001) DAMBE: software package for data analysis in molecular biology and evolution. J Hered 92:371–373

Xia X, Yang Q (2011) A distance-based least-square method for dating speciation events. Mol Phylogenet Evol 59:342–353

Xia X, Yang Q (2013) Cenancestor. In: Maloy S, Hughes K (eds) Brenner's encyclopedia of genetics. 2nd edition, Academic Press

Xia X, Yuen KY (2005) Differential selection and mutation between dsDNA and ssDNA phages shape the evolution of their genomic AT percentage. BMC Genet 6:20

Xia XH, Wei T, Xie Z, Danchin A (2002) Genomic changes in nucleotide and dinucleotide frequencies in *Pasteurella multocida* cultured under high temperature. Genetics 161:1385–1394

Yang MY, Bowmaker M, Reyes A, Vergani L, Angeli P, Gringeri E, Jacobs HT, Holt IJ (2002) Biased incorporation of ribonucleotides on the mitochondrial L-strand accounts for apparent strand-asymmetric DNA replication. Cell 111:495–505

Yasukawa T, Yang M-Y, Jacobs HT, Holt IJ (2005) A bidirectional origin of replication maps to the major noncoding region of human mitochondrial DNA. Mol Cell 18:651

Yu Q, Chen D, König R, Mariani R, Unutmaz D, Landau NR (2004) APOBEC3B and APOBEC3C are potent inhibitors of simian immunodeficiency virus replication. J Biol Chem 279:53379–53386

Zhang D, Xiong H, Shan J, Xia X, Trudeau V (2008) Functional insight into Maelstrom in the germline piRNA pathway: a unique domain homologous to the DnaQ-H 3'-5' exonuclease, its lineage-specific expansion/loss and evolutionarily active site switch. Biology Direct 3:48

Zilberstein D, Agmon V, Schuldiner S, Padan E (1982) The sodium/proton antiporter is part of the pH homeostasis mechanism in *Escherichia coli*. J Biol Chem 257:3687–3691

Zilberstein D, Padan E, Schuldiner S (1980) A single locus in *Escherichia coli* governs growth in alkaline pH and on carbon sources whose transport is sodium dependent. FEBS Lett 116:177–180

Index

X. Xia, *Comparative Genomics*, SpringerBriefs in Genetics,
DOI: 10.1007/978-3-642-37146-2, © The Author(s) 2013